He

F. Hall C.Ed., H.N.C., C.G.F.T.C.

Heating Technology

Second Edition

Longman Scientific & Technical

Longman Scientific & Technical,
Longman Group UK Limited,
Longman House, Burnt Mill, Harlow,
Essex CM20 2JE, England
and Associated Companies throughout the world.

First published 1981
Second edition 1986

Hall, F.
Heating technology. – 2nd ed.
1. Plumbing 2. Buildings – Great
Britain 3. Heating
I. Title
696'.12'0941 TH6122

ISBN 0-582-98877-2

Produced by Longman Group (FE) Limited
Printed in Hong Kong

To Nigel

Contents

Preface

The installation of hot and cold water supplies, fire control, gas, oil and heating systems forms an essential part of the work of the heating fitter. The heating fitter therefore requires a knowledge of the principles on which these systems are designed and has to install them so that they function efficiently and economically. During his studies at a technical college, the student heating fitter will follow a syllabus prepared by the City and Guilds of London Institute or the Regional Examining Bodies.

The primary aim of this book, therefore, is to assist the student heating fitter in his studies for the Advanced Craft Certificate, and the text covers closely the requirements of the syllabus in advanced technology. Although the book is written especially for the heating fitter craft student, it should also be useful to students taking the Chartered Institute of Building or the City and Guilds examination in Building Services and Equipment. Building surveying and architectural students taking the Technician Education Council's course should find the book useful for the unit in Building Services and Equipment. Craftsmen heating fitters and other personnel in the building industry should also find the book useful for reference purposes.

In preparing this book, constant reference has been made to relevant Byelaws, Regulations and Codes of Practice. There may be regional differences in the Water Byelaws and the local water authority should therefore be consulted before designing and installing a water supply system in any particular area. Wherever possible, the written descriptions have been amplified by illustrations, tables and calculations. I should like to thank the Publishers, Mr. C.R. Bassett for his encouragement and helpful criticisms, and Mrs. W.M. Whitney, librarian, for her help in obtaining various references from the Guildford County College of Technology, Department of Building and Surveying library. I should like to give very special thanks to my wife for her patience and understanding during the preparation of the book and for typing the entire manuscript.

F. Hall

Chapter 1
Water Supply

Rain Cycle

Water supply originates in nature in the form of rain, snow and hail falling from the clouds. Radiant heat from the sun causes evaporation of water on the earth's surface and the sea, thus forming clouds. The amount of water vapour that can be held by clouds depends on the temperature: when the temperature falls below the saturation point of the vapour, the clouds release the excess moisture, which falls to the earth. This process of evaporation and condensation is repeated and is known as the rain cycle.

Solvent Power of Water

Water has an extensive solvent capacity and there are very few substances that do not dissolve to some extent in water. Rainwater at the moment of formation is pure, but it soon ceases to be so. As rain falls through the atmosphere it dissolves some of the gases present, chiefly oxygen and carbon dioxide. It also washes the air so that the first fall is usually the most impure. By the time the water reaches the earth it contains various impurities and as it flows over or percolates through the soil it will dissolve other impurities present.

Sources of Water (Fig. 1.1)

Rainwater is soft and slightly acidic from the absorption of carbon dioxide (CO_2). On falling on moorland containing vegetable matter it dissolves more CO_2 and becomes distinctly, if weakly, acidic. This water will dissolve lead and is therefore known as plumbo-solvent. As the effects of lead poisoning are cumulative and very dangerous to health, such water must never be conveyed by lead pipes.

Shallow Wells. Shallow well water is obtained from sinkings in the top water-bearing strata of the earth. It should be treated with grave suspicion as it may become polluted from leaky drains or cesspools.

Intermittent or Land Springs. Since the water here is obtained from the same source as shallow well water it should be treated with the same grave suspicion.

Deep Wells. These are sinkings below the first impervious strata. Providing the well or borehole prevents the ingress of subsoil water, the water can usually be considered wholesome.

If the water passes through strata containing carbonate of calcium or magnesium, a certain amount of these salts will be dissolved in the water, depending on the amount of carbon dioxide present in the water. The water is now considered to be temporarily hard: this hardness can be removed by boiling the water, and can cause scaling of hot water pipes and boilers.

If the water passes through strata containing calcium sulphate, calcium chloride or

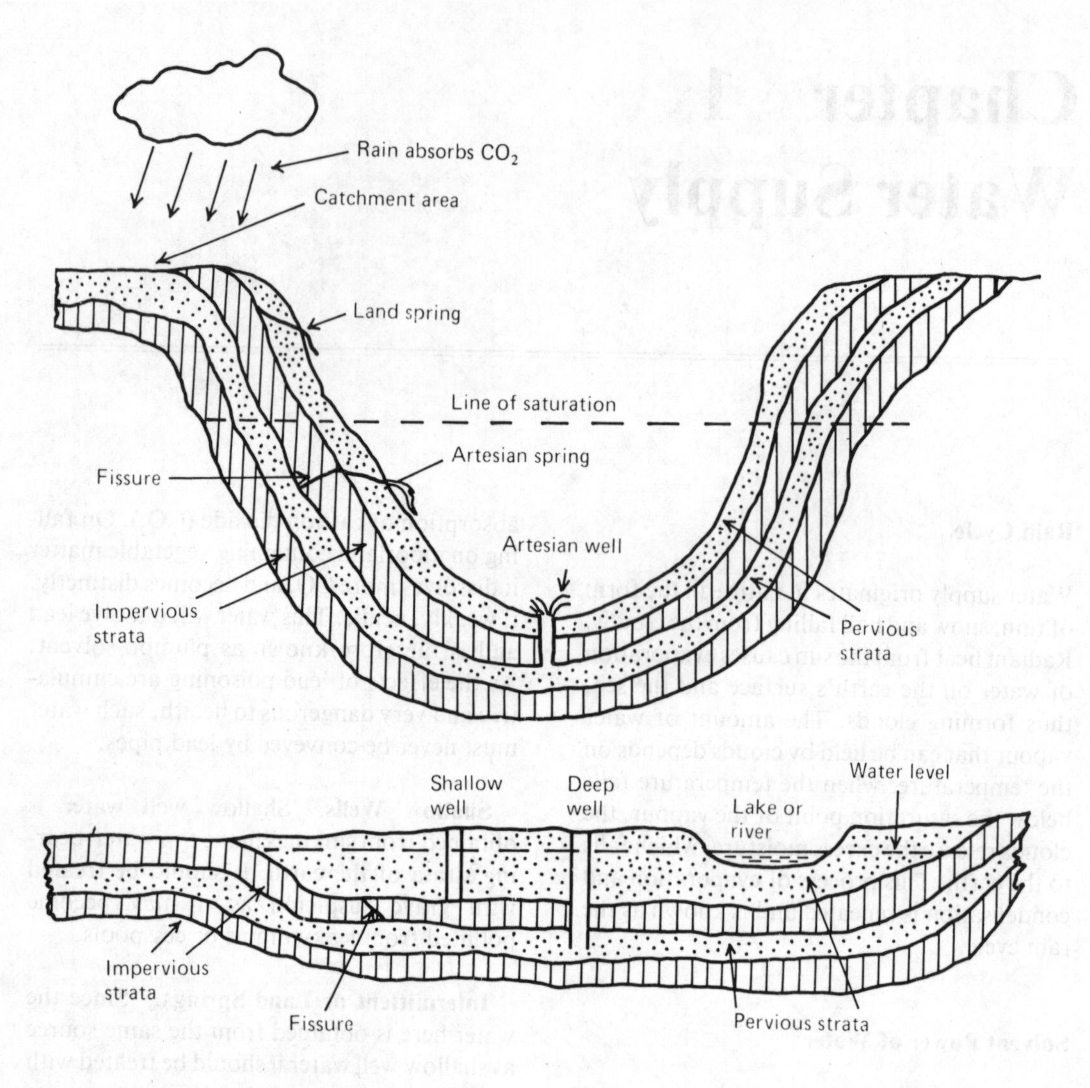

Fig. 1.1 Sources of water supply.

magnesium chloride, a certain amount of these salts will be dissolved in the water without the presence of carbon dioxide. This type of hardness cannot be removed by boiling the water and is termed permanent hardness. It will not cause scaling of hot water pipes and boilers, but it may cause corrosion.

Most waters contain both temporary and permanent hardness. The generally accepted classification of hardness is given in Table 1.1.

Table 1.1

Type of water	Hardness in parts per million
Soft	0 – 50
Moderately soft	50 – 100
Slightly hard	100 – 150
Moderately hard	150 – 200
Hard	200 – 300
Very hard	over 300

Artesian Wells and Springs. The water is obtained from the same source as deep well water and can be considered wholesome.

Lakes and Rivers. These are formed chiefly from the catchment of surface and

Table 1.2

Wholesome	1	Spring water	Very palatable
	2	Deep well water	
	3	Upland surface water	Moderately palatable
Suspicious	4	Stored rainwater	
	5	Surface water from cultivated lands	Palatable
Dangerous	6	River water to which sewage gains access	
	7	Shallow well water	

subsoil water. Water impounded in lakes from upland surfaces is soft and usually wholesome. River water is soft and generally turbid, especially after a storm: it may be rendered dangerous by discharges from factories and sewage works.

Classification of Water from Various Sources

The River Pollution Commissioners classify water from the various sources as shown in Table 1.2.

Analysis of Water

Two examinations should be made, namely for (a) bacteriological and (b) chemical content. The former is the more important from the health aspect, but the chemical content of the water must be considered because of its relation to attack on metals. If a new source of supply is to be used for

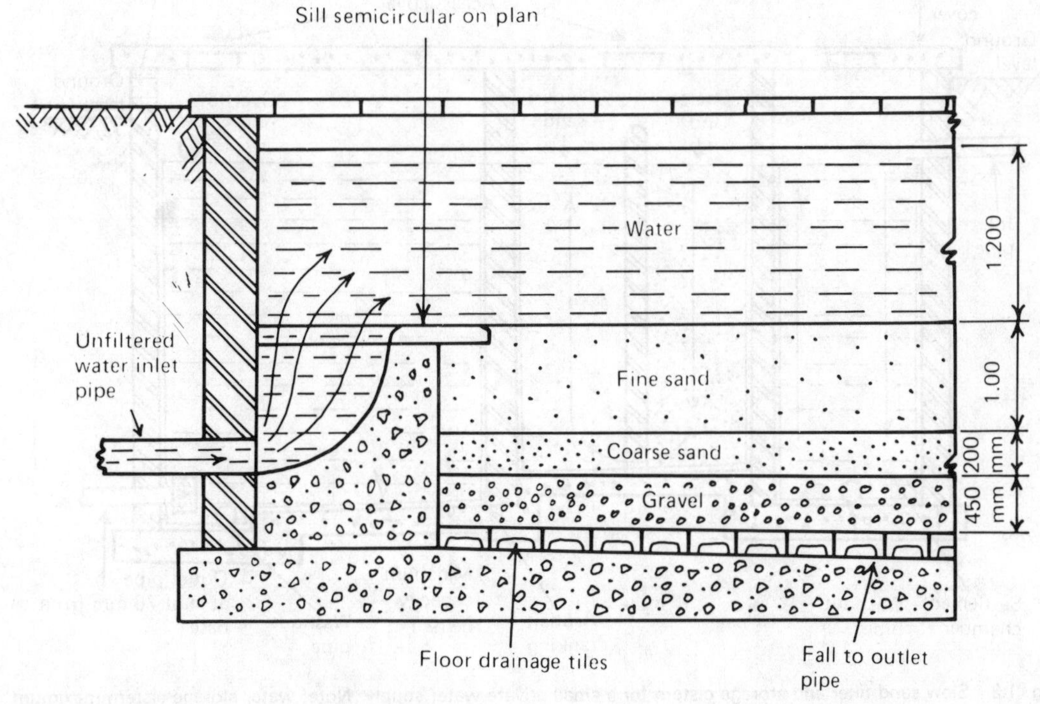

Fig. 1.2 Longitudinal section of a slow sand filter.

domestic purposes, or where doubt exists with an existing supply, the water should be submitted to an analyst for examination and report.

Filtration of Water

Slow Sand Filters. The simplest form of filter is one where the water passes downward by gravity through layers of sand and gravel (see Fig. 1.2). When the filter is first used it acts simply as a scourer, removing suspended matter but not removing harmful bacteria. In time, however, colloidal matter forms in the interstices between the sand grains. This gelatinous film is a barrier to the passage of harmful bacteria but gradually slows the passage of water to such an extent that it becomes necessary to remove the accumulated sludge. A certain amount of sand is removed in the process and it is therefore essential periodically to replace the surface with new sand. Fig. 1.3 shows a filter and storage cistern suitable for a small private water supply.

Pressure Filters. These consist of steel cylinders with the bottom filled with gravel and the remainder with sand (see Fig. 1.4). Water enters at the top and is collected in a perforated plate at the bottom, where there is a connection to an outlet pipe. The principle of operation is the same as the gravitational or slow sand filter, but since the water entering is under pressure the filtering process is much quicker. The efficiency of the filter is increased by adding a small dose of aluminium sulphate to the inlet water, which then forms a gelatinous film on top of the sand. The sand is cleaned by back washing and scouring with compressed air. The cylinder may be up to 2.7 m in diameter and the rate of filtration up to 12 m³ per m² of horizontal surface per hour.

Domestic Filters. The filter consists of an

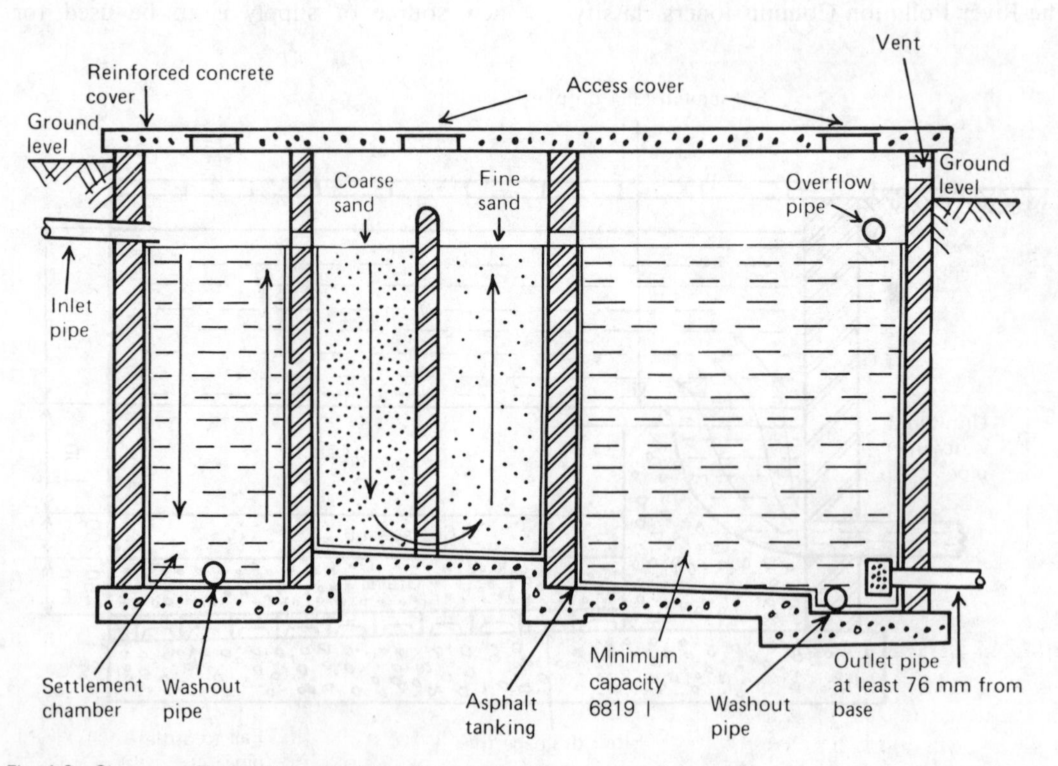

Fig. 1.3 Slow sand filter and storage cistern for a small private water supply. Note: water storage cistern maximum capacity 6820 l.

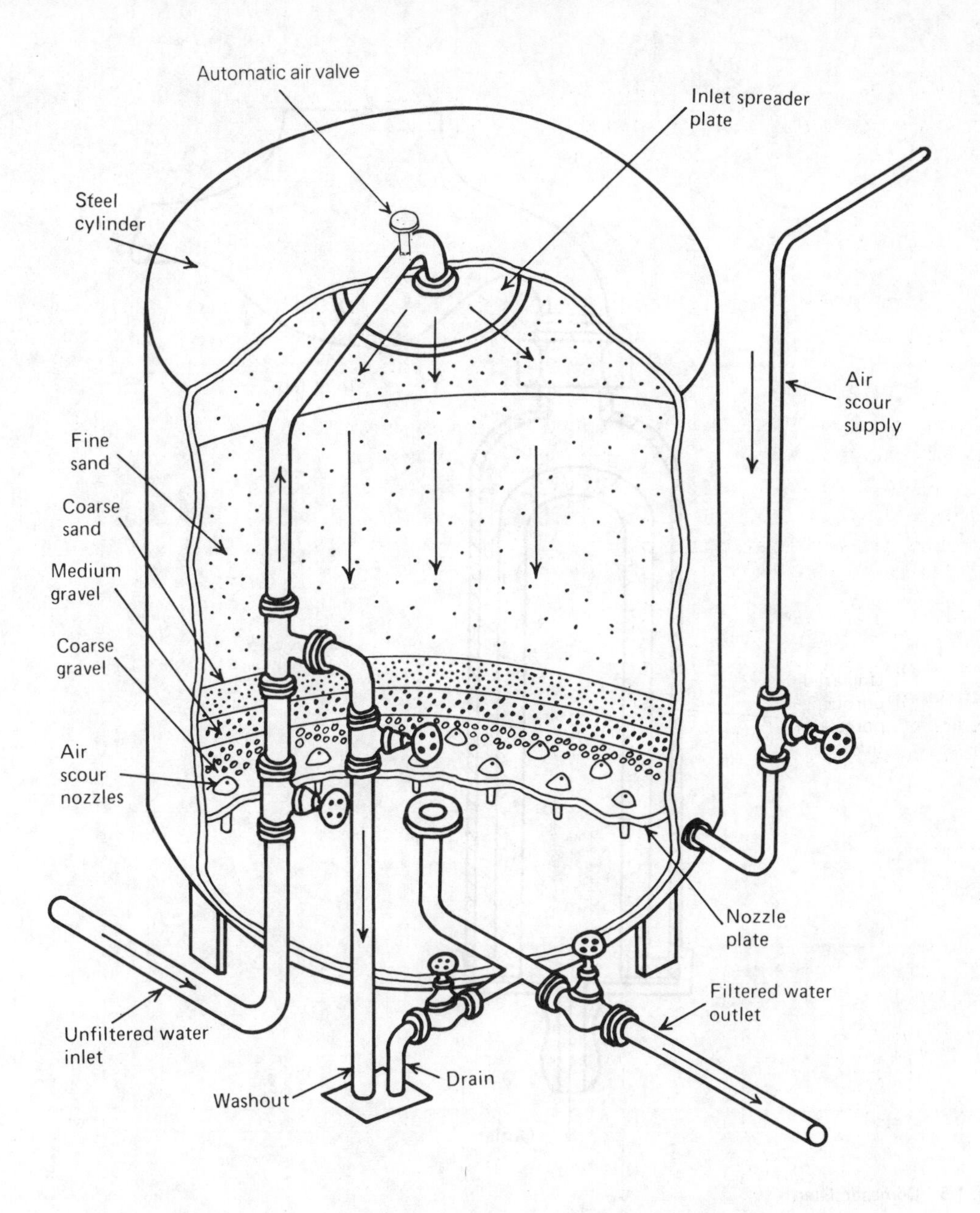

Fig. 1.4 Sectional view of Candy vertical-type pressure filter.

unglazed porous porcelain cylinder through which the water to be filtered flows. The filter is periodically cleaned in boiling water and impregnated with silver nitrate solution which has a sterilising effect upon the water. For passing larger volumes of water, filters may be obtained in batteries enclosed inside a cylinder. The type shown in Fig. 1.5 may be attached to any drinking water outlet tap.

Sterilisation of Water

In order to render large quantities of water safe for human consumption, sterilisation is required to destroy harmful bacteria.

Chlorine, because of its great efficiency when used in small quantities, is the most common reagent for the sterilisation of water. Its germicidal action in small doses is

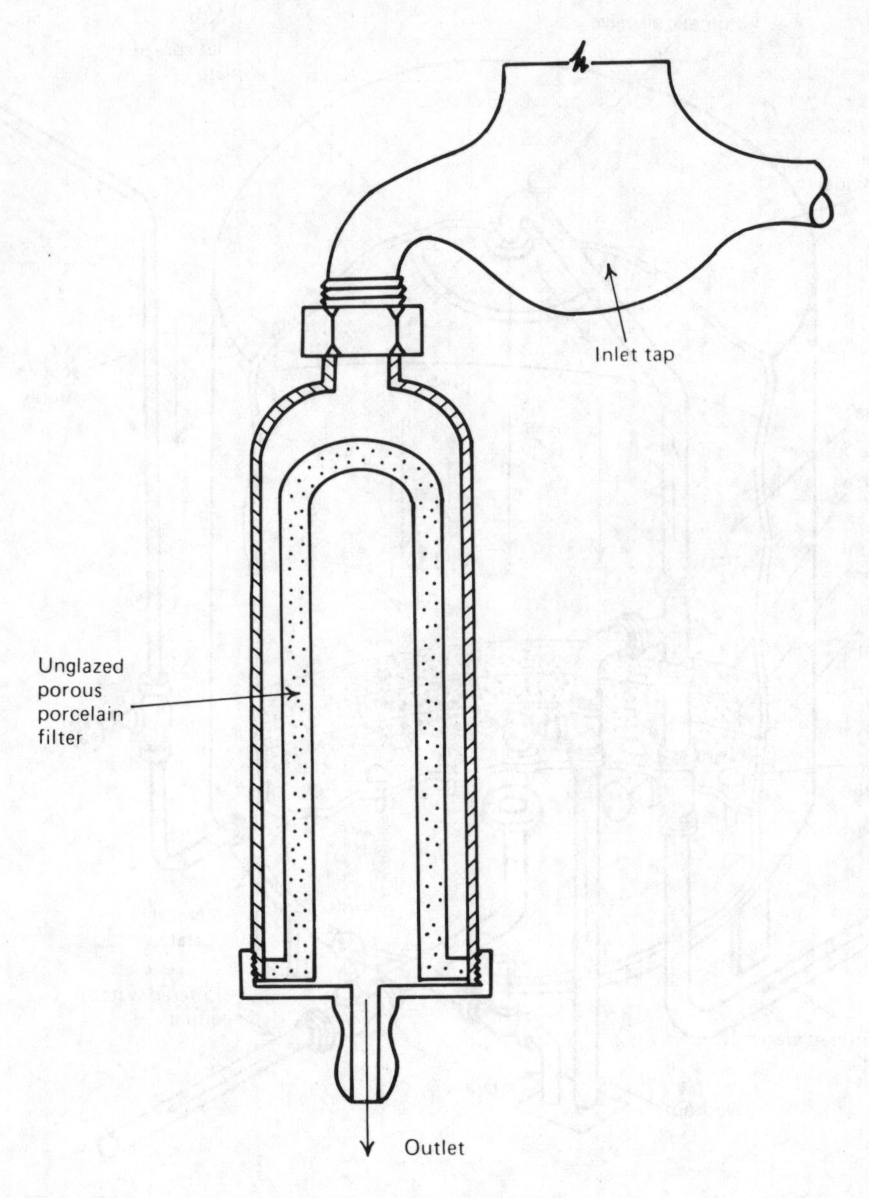

Fig. 1.5 Domestic filter.

due to the destruction of enzymes necessary for the existence of microorganisms. It also has considerable oxidising powers, which favour the destruction of organic matter. The dosage of chlorine is strictly regulated so that there is sufficient to destroy any bacteria present but not enough to give an unpleasant taste to the water.

The chlorine is stored as a gas in steel cylinders from which it is injected into the water by automatic equipment. Fig. 1.6 shows details of a chlorinating plant which will automatically inject the correct amount of chlorine into a water main.

Water Softening

Hard waters are objectionable in domestic installations because more soap is required to produce a lather than when soft water is used. The term 'hardness' refers to the difficulty of

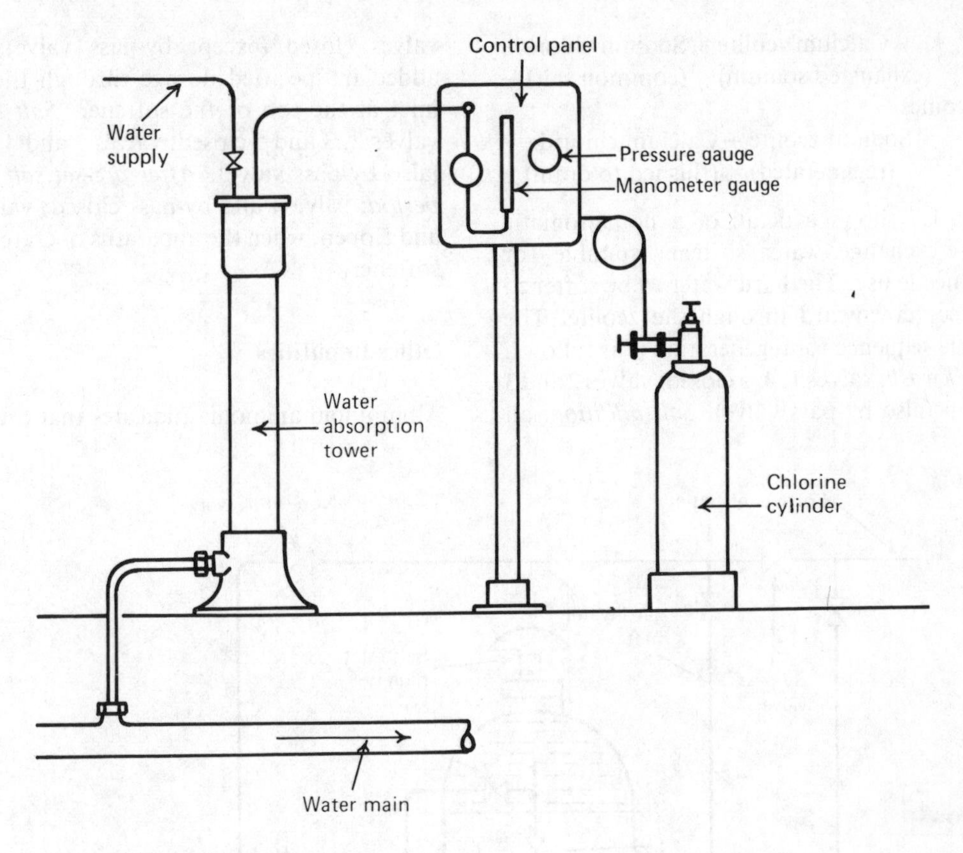

Fig. 1.6 Detail of chlorinating plant.

obtaining a lather with soap.

Permanently hard water may be softened by the use of sodium carbonate (washing soda), which causes the precipitation of calcium carbonate, leaving soluble sodium sulphate in solution.

Temporarily hard water may be softened by the use of slaked lime, which takes up the carbon dioxide from the bicarbonate present in the water, resulting in the precipitation of insoluble carbonate and the removal of temporary hardness. Slaked lime is used in conjuction with sodium carbonate in what is known as the *lime-soda process* of water softening.

Base-Exchange Process. This process is used extensively for both industrial and domestic installations. The process removes both temporary and permanent hardness by passing the hard water through sodium zeolites. Sodium zeolites have the property of exchanging their sodium base for the calcium or magnesium base, hence the term 'base exchange'. The process is as follows:

	Sodium zeolite +	Calcium carbonate or sulphate
	(inside softener)	(in water)
becomes		
	Calcium zeolite +	Sodium carbonate or sulphate
	(held inside softener)	(in solution with the water)

After a period of use, the sodium zeolite is converted into magnesium and calcium zeolite, which has no softening power. It is therefore *regenerated* by the addition of common salt (sodium chloride). The salt is kept in contact with the magnesium and calcium zeolite for about half an hour, by which time the original sodium zeolite is produced. The process is as follows:

Calcium zeolite + Sodium chloride
(exhausted sodium) (common salt)

becomes

Sodium zeolite + Calcium chloride
(regenerated) (flushed to drain)

Fig. 1.7 shows a detail of a nonautomatic base-exchange water softener suitable for domestic use. The hard water to be softened passes downward through the zeolite. The value sequence for regeneration is as follows. *Backwash*: valves 1, 4, 5 closed; valves 2 and 3 open (also by-pass valve). *Salt addition*: all valves closed (except by-pass valve); salt added in specified dosage through the salt inlet at the top of the softener. *Salt rinse*: valves 2, 3 and 5 closed; valves 1 and 4 open (also by-pass valve). *After pre-set salt rinse period*: valve 4 and by-pass closed; valves 1 and 5 open, when the apparatus operates as a softener.

Other Impurities

Albuminoid ammonia indicates that organic

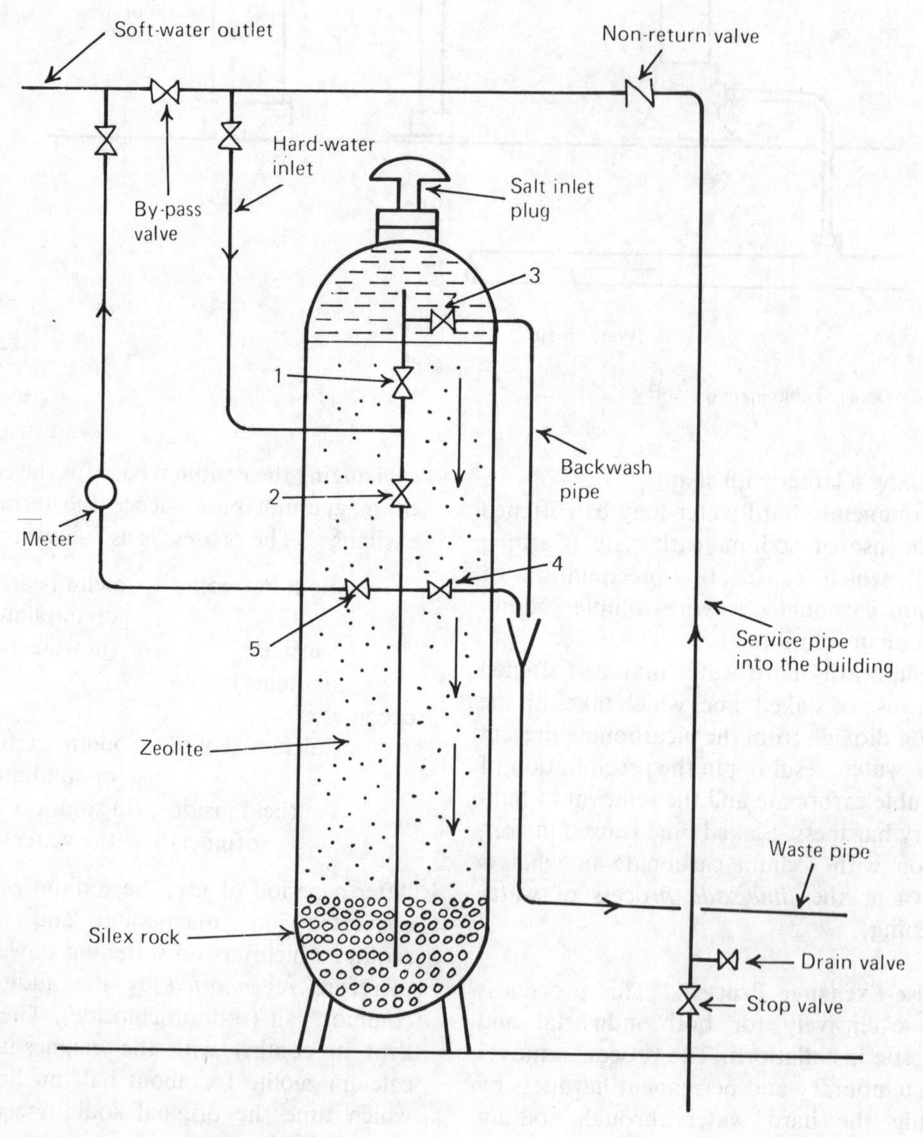

Fig. 1.7 Base-exchange water softener.

matter exists that is still undecomposed and gives the water a disagreeable taste and odour. Nitrites are a special danger as they indicate the presence of organic pollution. Nitrites represent the transitional stage in the oxidation of organic matter between ammonia and nitrates.

Nitrates signify past pollution and their presence without nitrites is an indication that the organic matter has been completely oxidised.

Water Quality

Water used for human consumption must be free from harmful bacteria and suspended matter, colourless, pleasant to taste, and for health reasons moderately hard.

Exercises

1. Sketch and describe the rain cycle.
2. Define
   (a) shallow well,
   (b) deep well,
   (c) artesian well,
   (d) spring.
3. Describe the characteristics of water from various supply sources.
4. Define
   (a) soft water,
   (b) temporary hardness of water,
   (c) permanent hardness of water.
5. State the two examinations required for water and explain the significance of each.
6. Sketch and describe
   (a) a slow sand filter,
   (b) a pressure filter,
   (c) a domestic filter.
7. Describe how water is sterilised.
8. Describe three methods of softening hard water.
9. Sketch a base-exchange water softener and describe its action.
10. What is the significance of the presence of albuminoid ammonia, nitrates and nitrates in water?
11. What are the characteristics of water for human consumption?
12. Define 'hardness' of water and describe how rainwater becomes 'hard'.
13. Sketch a section through an underground water storage tank.
14. State the advantages and disadvantages of slow sand filters and rapid filters.

Chapter 2
Cold Water Supply Systems

Definitions

The following definitions are used by most water authorities.

CISTERN: a container for water having a free water surface at atmospheric pressure.

FEED CISTERN: any storage cistern used for supplying cold water to a hot water apparatus.

STORAGE CISTERN: any cistern, other than a flushing cistern, having a free water surface under atmospheric pressure, but not including a drinking trough or drinking bowl for animals.

CAPACITY (of a cistern): the capacity up to the water line.

WATER LINE: a line marked inside a cistern to indicate the water level at which the ballvalve should be adjusted to shut off.

OVERFLOWING LEVEL: in relation to a warning or other overflow pipe of a cistern, the lowest level at which water can flow into that pipe from a cistern.

WARNING PIPE: an overflow pipe so fixed that its outlet end is in an exposed and conspicuous position and where the discharge of any water from the pipe may be readily seen and, where practicable, outside the building.

COMMUNICATION PIPE: any service pipe from the water main to the stop valve fitted on the pipe.

SERVICE PIPE: so much of any pipe for supplying water from a main to any premises as is subject to water pressure from that main, or would be so subject but for the closing of some stop valve.

DISTRIBUTING PIPE: any pipe for conveying water from a cistern, and under pressure from that cistern.

SUPPLY PIPE: so much of any service pipe which is not a communicating pipe.

MAIN: a pipe for the general conveyance of water as distinct from the conveyance to individual premises.

HOT WATER CYLINDER OR TANK: a closed container for hot water under more than atmospheric pressure. Note: a cylinder is deemed to include a tank.

POTABLE: water suitable for drinking.

FITTING: anything fitted or fixed in connection with the supply, measurement, control, distribution, utilisation or disposal of water.

Connection to Water Main

Before any building can be supplied with water from the main, it is essential to provide adequate notice in writing to the local water authority.

The tapping of the main and laying of the communication pipe is usually carried out by the local water authority at the building owner's expense. Where the local water authority permits a contractor to lay the communication pipe, the connection to the main will usually be made by the authority, also at the building owner's expense. Any underground piping should be inspected by the local water authority before being filled in.

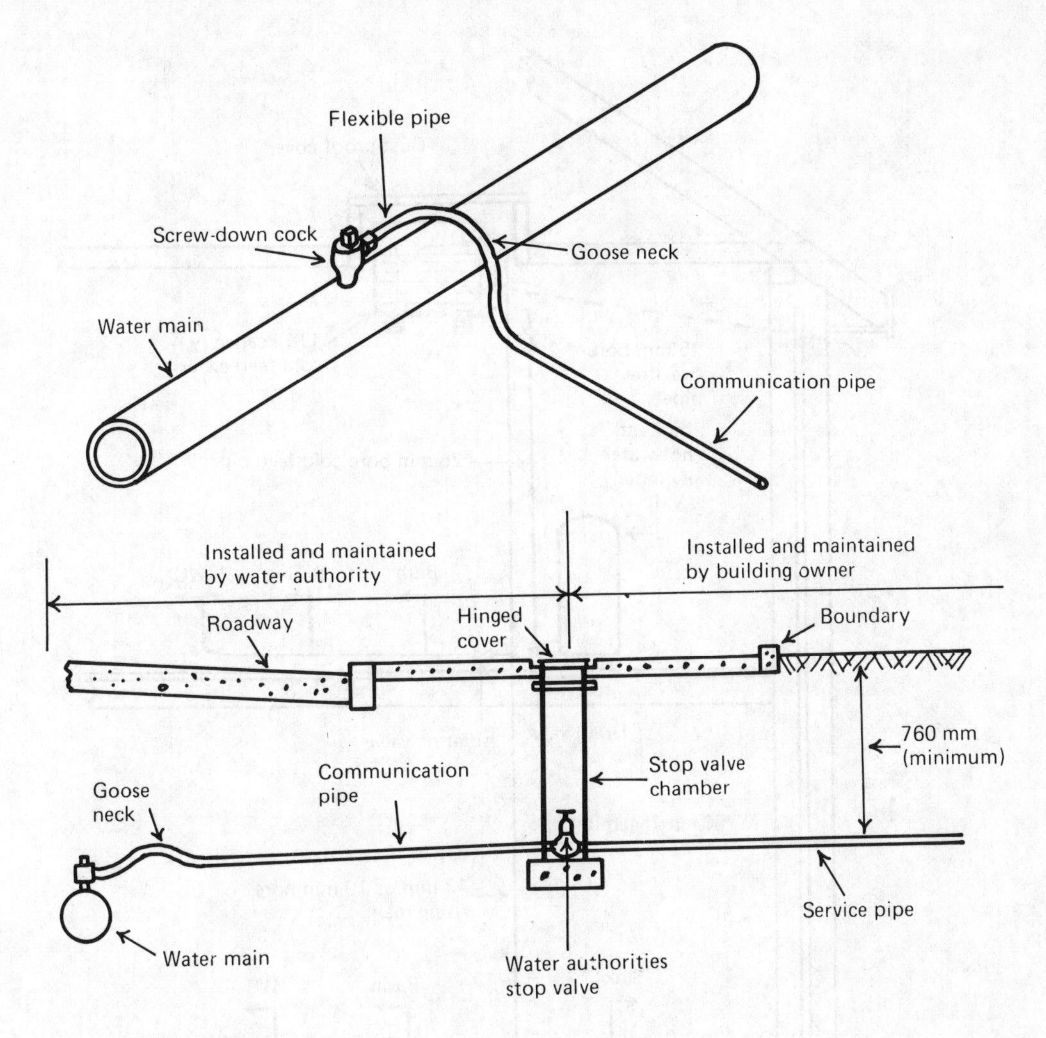

Fig. 2.1 Connection to water main.

In order to allow for any settlement of the communication pipe, a bend is made where the pipe connects to the main. Fig. 2.1 shows how the communication pipe is connected to the main and laid below ground.

Distribution Systems

Two distinct types of cold water system are used, depending on the water authority regulations.

Nonstorage or Direct Systems. In nonstorage or direct systems, all the sanitary fittings are supplied with cold water direct from the main. A cold water feed cistern is usually required to 'feed' the hot water supply system.

With certain types of electric or gas water heaters that may be supplied direct from the main, a cold water feed cistern is not required and this simplifies the system.

Fig. 2.2 shows a direct or nonstorage system for a house. The cold water feed cistern is small enough to be housed in the top of the airing cupboard, thus avoiding the risk of freezing.

Fig. 2.3 shows a direct or nonstorage system for a three-storey office or factory. The open outlet type gas or electric water heaters, which are connected direct from the main, avoid the use of a cold water feed cistern. A swivel outlet from the heaters may

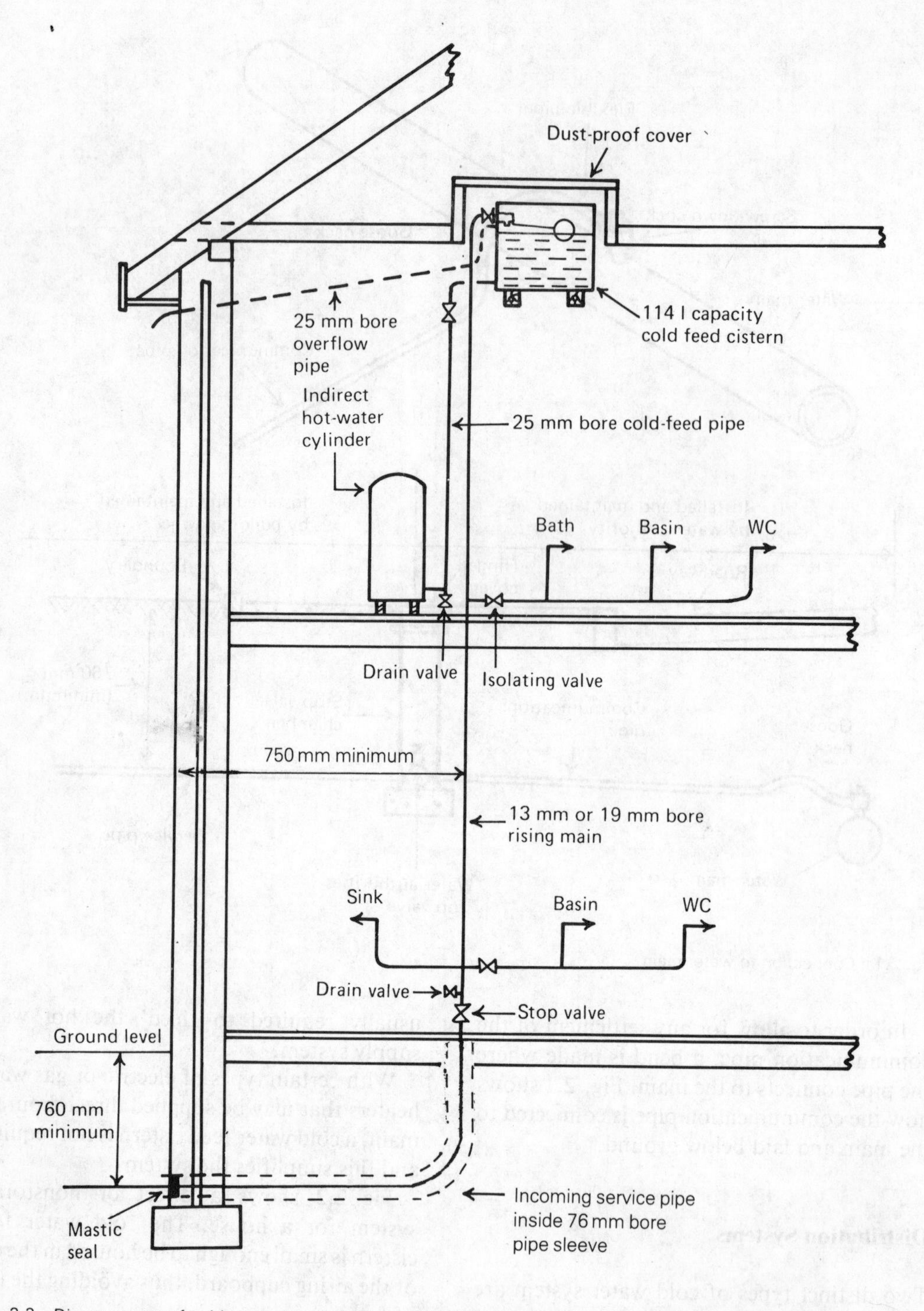

Fig. 2.2 Direct system of cold water supply for a house.

be turned to supply hot water to each wash basin.

Indirect or Storage Systems. In indirect or storage systems all the drinking water used in the building is supplied from the main and water used for all other purposes is supplied indirectly from a cold water storage cistern.

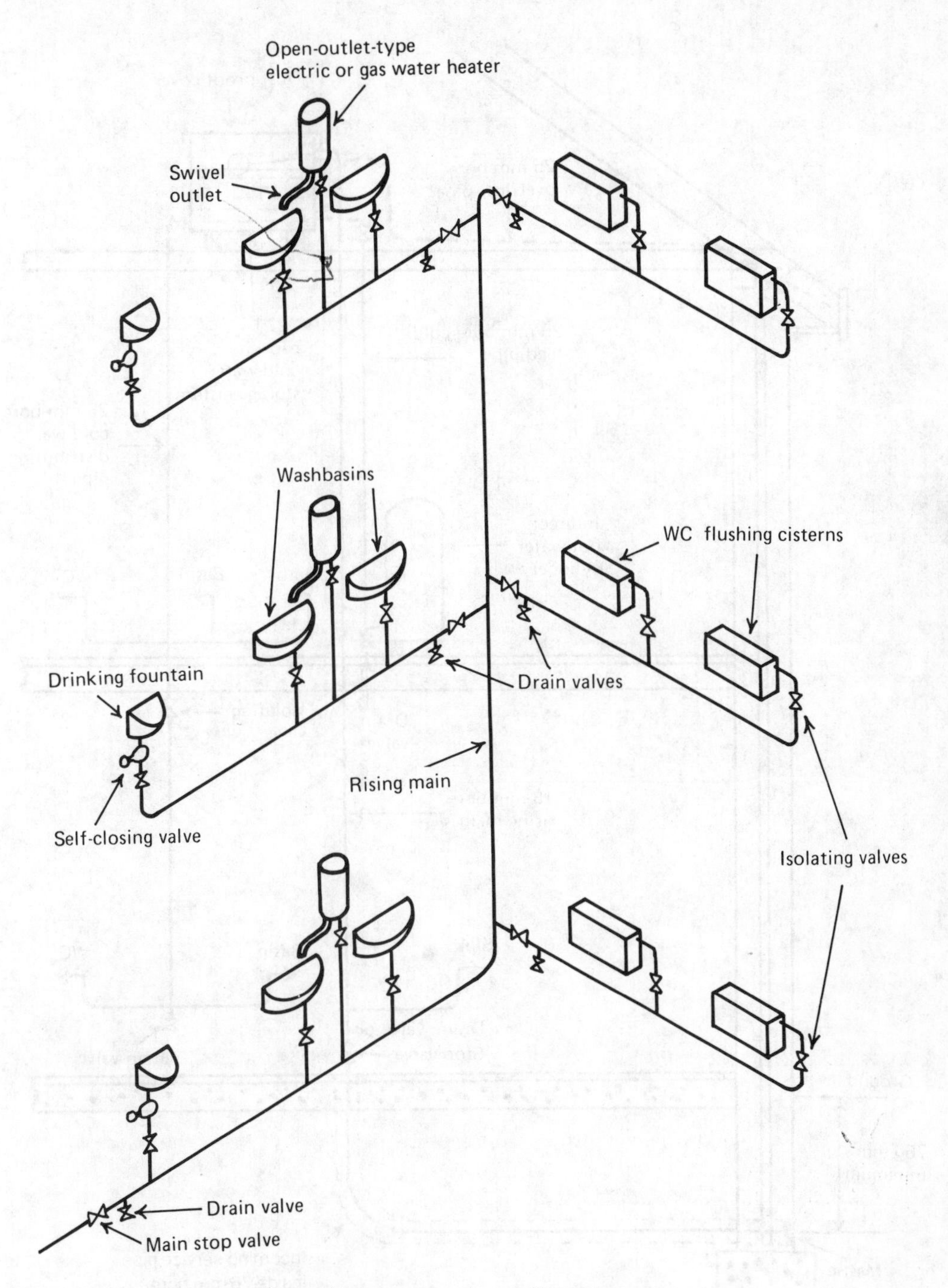

Fig. 2.3 Isometric diagram of a direct cold water system for an office or factory.

Since the cistern also supplies water to the hot water cylinder its capacity will be almost double the capacity required for the direct system.

Fig. 2.4 shows an indirect or storage system for a house. The cold water storage cistern is too large to be housed in the top of the airing cupboard and is therefore housed in the roof space near to the chimney breast to prevent freezing.

Fig. 2.5 shows an indirect or storage system for a three-storey office of factory.

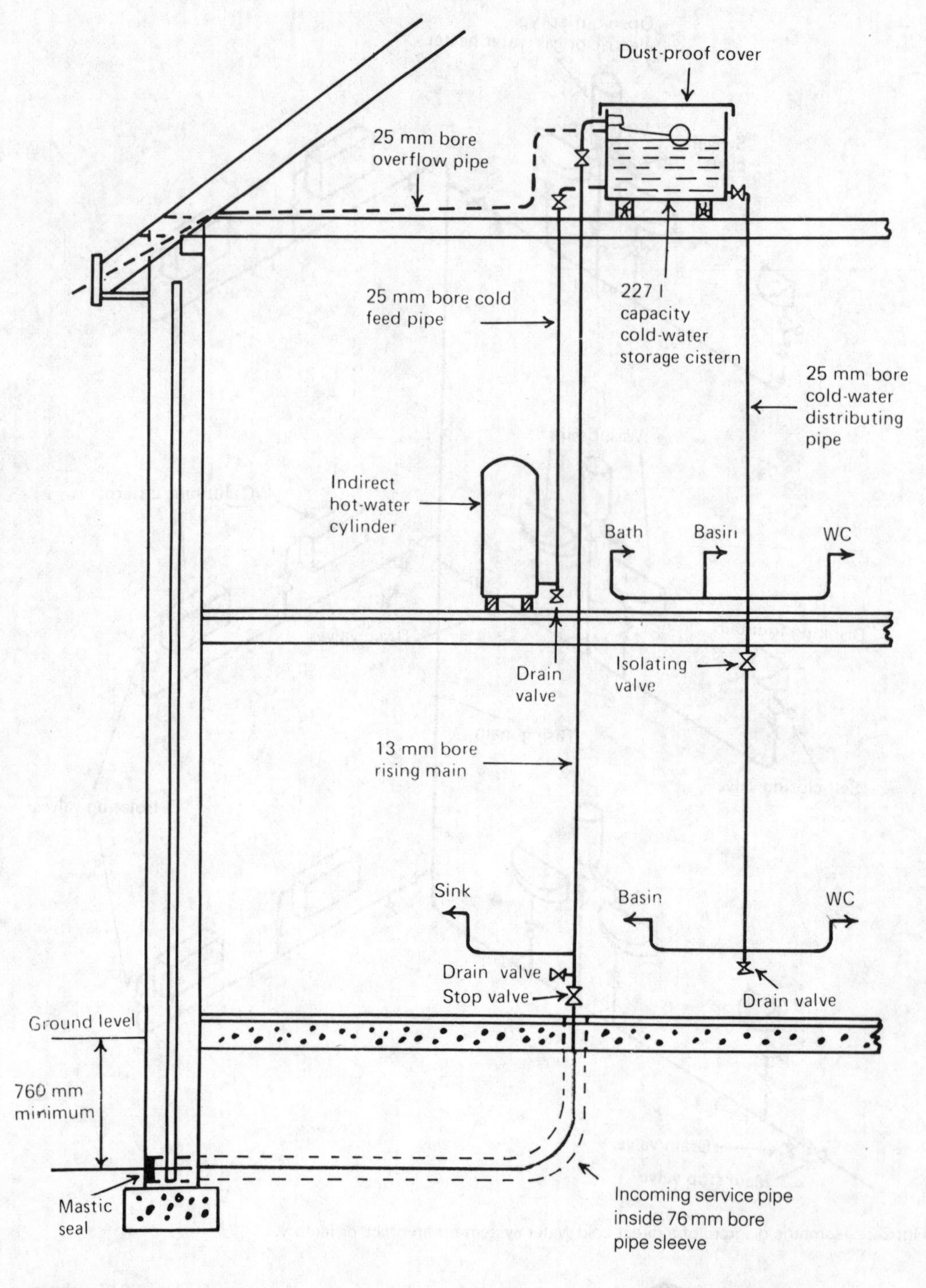

Fig. 2.4 Indirect system of cold water supply for a house.

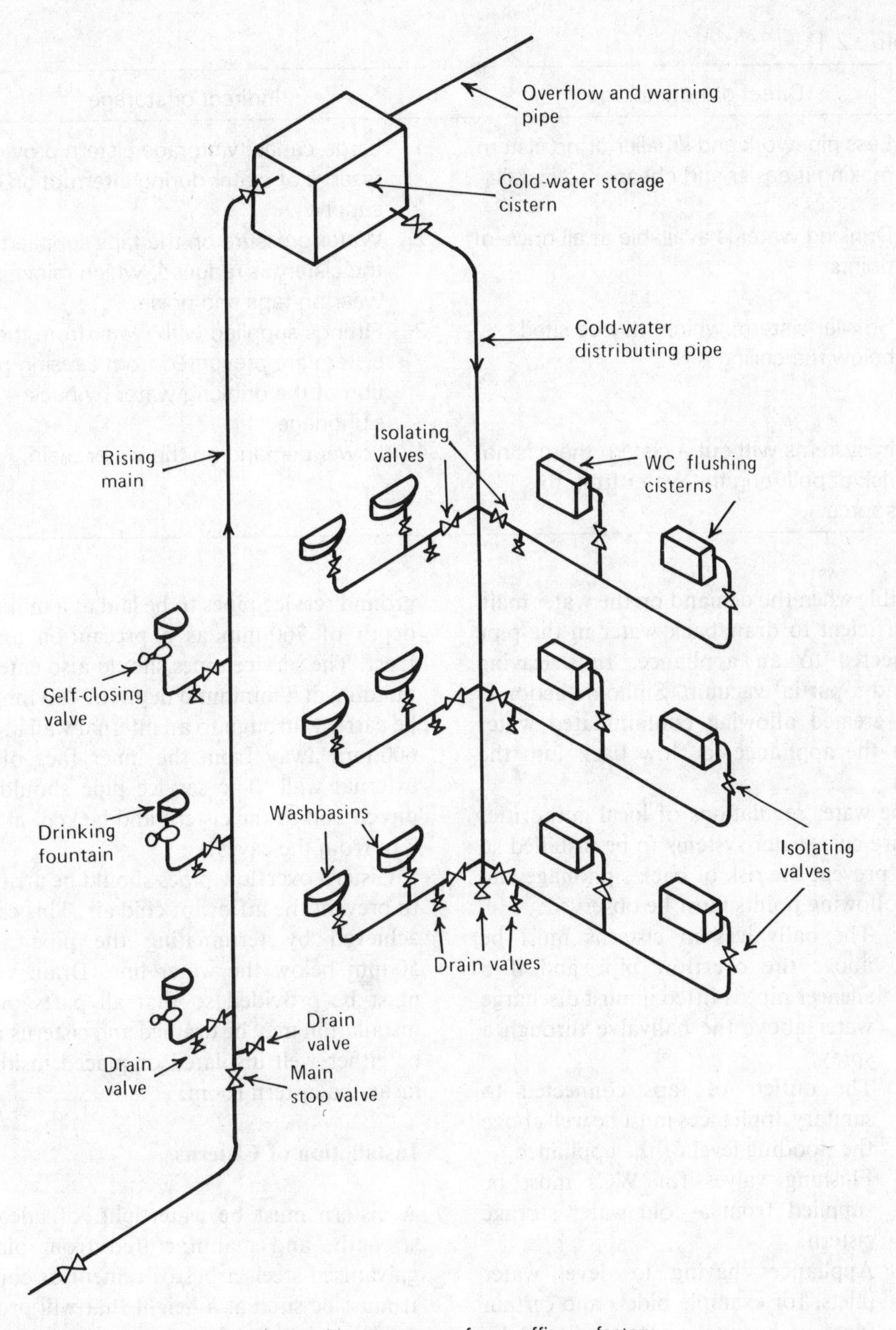

Fig. 2.5 Isometric diagram of an indirect cold water system for an office or factory

Advantages of direct and indirect cold water systems are given in Table 2.1.

Prevention of Back Siphonage

Back siphonage is the back flow of water, which may be contaminated, into the drinking water supply. In order for back siphonage to occur, a negative pressure or partial vacuum must be created in the pipe connected to an appliance having its outlet submerged in water, which may be contaminated. This is

Table 2.1

Direct or nonstorage	Indirect or storage
1. Less pipework and smaller or no cistern, making it easier and cheaper to install.	1. Large-capacity storage cistern provides a reserve of water during interruption of supply.
2. Drinking water is available at all draw-off points.	2. Water pressure on the taps supplied from the cistern is reduced, which minimises wear on taps and noise.
3. Smaller cistern, which may be sited below the ceiling.	3. Fittings supplied with water from the cistern are prevented from causing pollution of the drinking water by back siphonage.
4. In systems without a cistern there is no risk of polluting the water from this source.	4. Lower demand on the water main.

possible when the demand on the water main is sufficient to draw back water in the pipe connected to an appliance, thus leaving behind a partial vacuum. Siphonic action is then created allowing contaminated water from the appliance to flow back into the main.

The water regulations of local authorities require cold water systems to be installed so as to prevent the risk of back siphonage and the following points must be observed:

1. The ballvalves in cisterns must be above the overflow pipe and if a silencer pipe is fitted it must discharge water above the ballvalve through a spray.
2. The outlets of taps connected to sanitary appliances must be well above the flooding level of the appliance.
3. Flushing valves for WCs must be supplied from a cold water storage cistern.
4. Appliances having low-level water inlets, for example bidets and certain types of hospital appliance, must be supplied from a cold water storage cistern and never direct from the main.

Precautions Against Frost Damage

Water regulations generally require underground service pipes to be laid at a minimum depth of 760 mm as a precaution against frost. The service pipes should also enter the building at a minimum depth of 760 mm and be carried through to an internal wall at least 600 mm away from the inner face of any external wall. The service pipe should run directly up to the cistern and be kept at least 2 m from the eaves.

Cistern overflow pipes should be arranged to prevent the inflow of cold air. This can be achieved by terminating the pipe about 50 mm below the water line. Drain valves must be provided so that all parts of the installation may be drained and cisterns must be either well insulated or placed inside an insulated cistern room.

Installation of Cisterns

A cistern must be watertight, of adequate strength, and manufactured from plastic, galvanised steel, asbestos cement or copper. It must be sited at a height that will provide sufficient head and discharge of water to the fittings supplied. It must be placed in a position where it can be readily inspected and cleansed. A cistern must be provided with a dust-proof but not airtight cover and protected from damage by frost.

Every storage cistern must be fitted with an efficient overflow pipe which should have a fall as great as practicable not less than 1 in 10. In England and Wales the regulations require the overflow pipe to have an internal diameter greater than the inlet pipe and in no case less than 19 mm. In Scotland the regulations require an overflow pipe to have an internal diameter of not less than twice that of the inlet pipe and in no case less than 32 mm.

If the capacity of a storage cistern does not exceed 4546 litre, the overflow pipe should be arranged as a warning pipe, i.e. so that its outlet is in a conspicuous position, either inside or outside the building, so that any discharge of water can be readily seen. There should be no other overflow pipe. If the capacity of a storage cistern exceeds 4546 litre it should have a warning pipe as previously described, or alternatively it should have an overflow pipe not arranged as a warning pipe and, in addition, a warning pipe of not less than 25 mm internal diameter, or some other device, which effectively indicates when the water reaches a level not less than 50 mm below the overflowing level of the overflow pipe.

Fig. 2.6 shows the method of installing a cistern in a roof space.

Duplication of Cisterns CP 310: *Water Supply* recommends that if the storage required exceeds 4500 litre it is an advantage to provide two or more cisterns interconnected so that each cistern may be cleansed or renewed without cutting off the supply of water from the remaining cisterns. Fig. 2.7

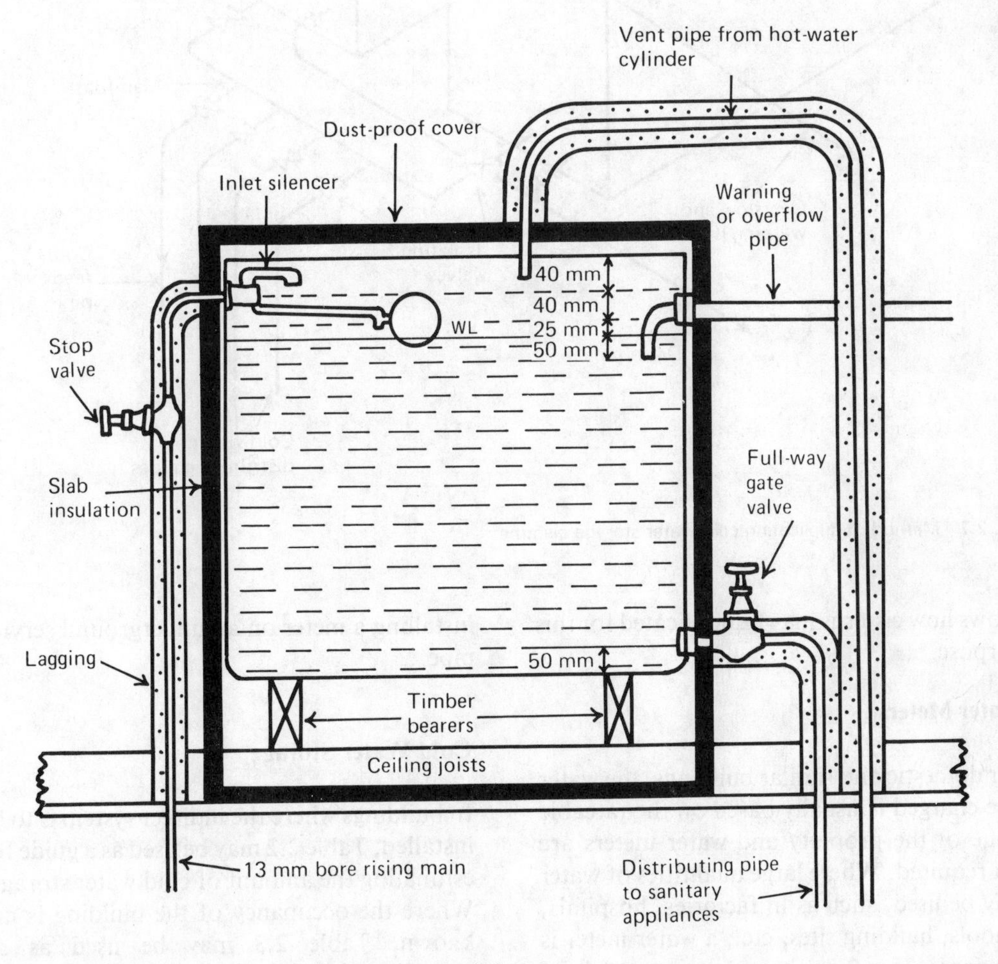

Fig. 2.6 Method of installing cold water storage or feed cistern.

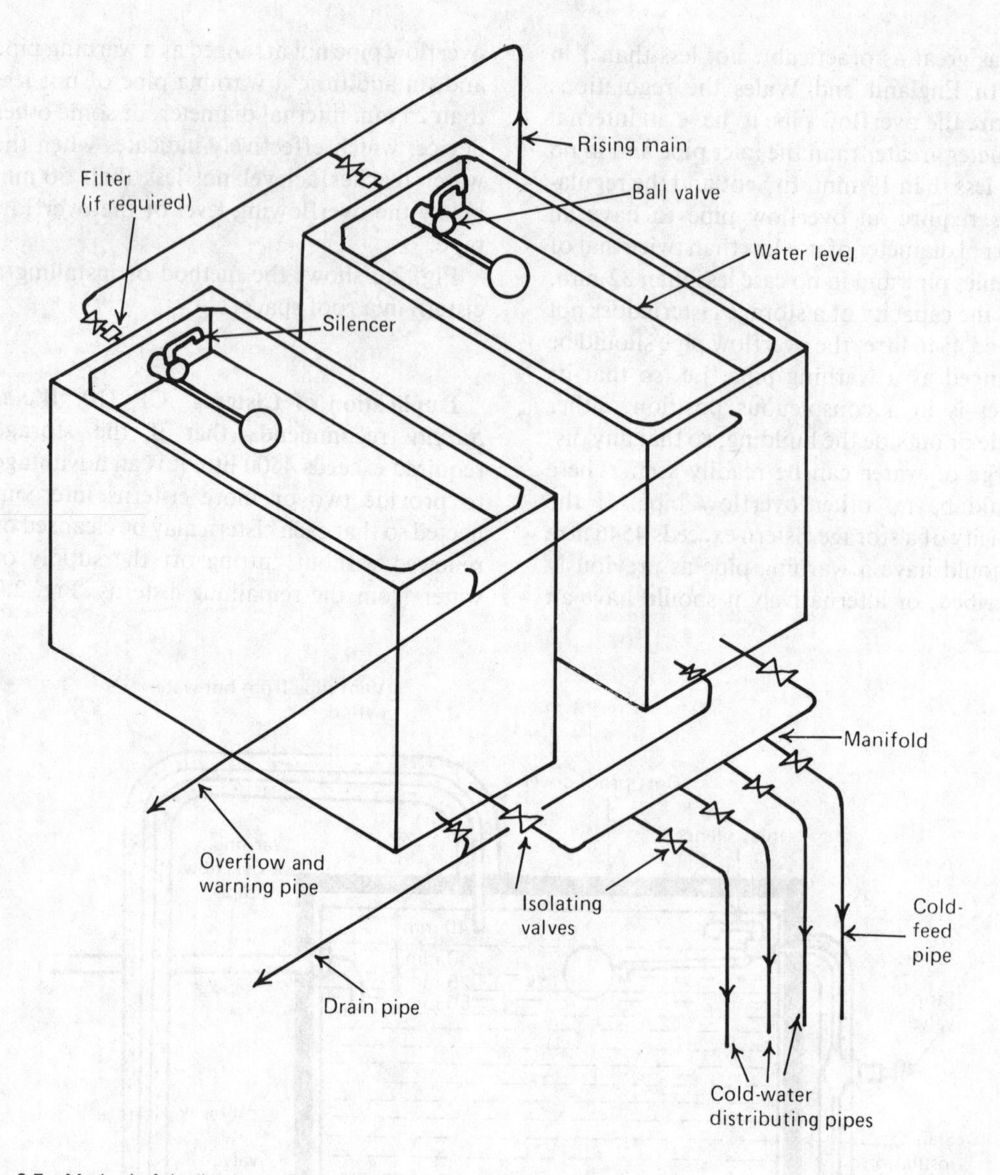

Fig. 2.7 Method of duplicating cold water storage cisterns.

shows how cisterns may be duplicated for this purpose.

Water Meters

For domestic and similar buildings, the water rate charged is usually based on the rateable value of the property and water meters are not required. Where large quantities of water may be used, such as in factories, hospitals, schools, building sites, etc., a water meter is required. Fig. 2.8 shows the method of installing a meter on an underground service pipe.

Cold Water Storage

In buildings where the indirect system is to be installed, Table 2.2 may be used as a guide for estimating the amount of cold water storage. Where the occupancy of the building is not known, Table 2.3 may be used as an approximate guide to the storage required.

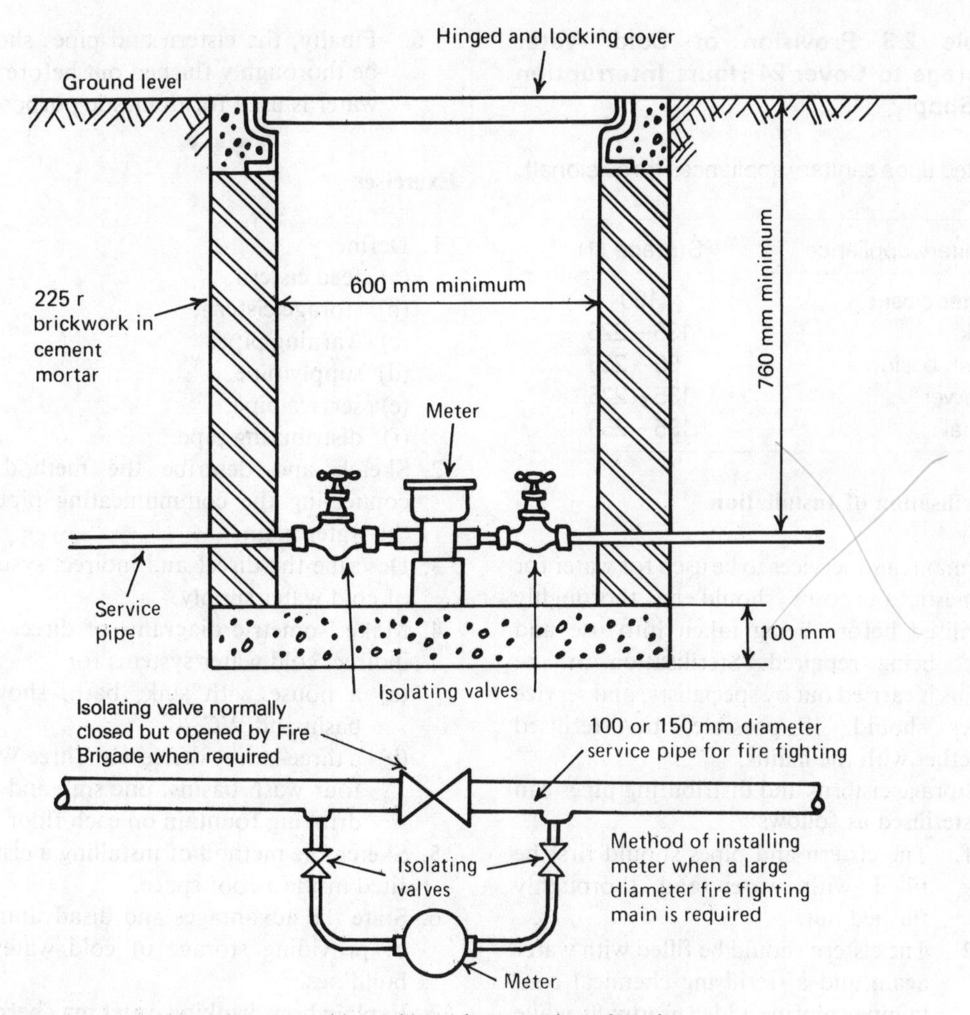

Fig. 2.8 Method of installing a meter on the underground service pipe.

Table 2.2 Provision of Cold Water Storage to Cover 24 Hours Interruption of Supply

Based upon occupancy, reproduced by permission of the British Standards Institution.

Type of building		Storage (1)	Type of building		Storage (1)
Dwelling houses and flats	per resident	90	Offices with canteens	per head	45
Hostels	per resident	90	Restaurants	per head/per meal	10
Hotels	per resident	140	Day schools	per head	30
Offices without canteens	per head	40	Boarding schools	per head	90
			Nurses' homes and medical quarters	per resident	115

Table 2.3 Provision of Cold Water Storage to Cover 24 Hours Interruption of Supply

Based upon sanitary appliances (provisional).

Sanitary appliance	Storage (1)
Water closet	180
Sink	135 – 225
Wash basin	90 – 250
Shower	135 – 225
Urinal	135 – 250

Sterilisation of Installation

All mains and services to be used for water for domestic purposes should be thoroughly sterilised before being taken into use and after being repaired. Sterilisation of the mains is carried out by specialists, and service pipes should, if possible, be sterilised together with the mains.

Storage cisterns and distributing pipes can be sterilised as follows:

1. The cistern and pipes should first be filled with water and thoroughly flushed out.
2. The cistern should be filled with water again and a sterilising chemical containing chlorine added gradually while the cistern is filling. Sufficient chemical should be added to give the water a dose of 50 parts of chlorine to one million parts of water.
3. When the cistern is full, the supply should be stopped, and all the taps on the distributing pipes opened successively, working progressively away from the cistern. Each tap should be closed when water discharged smells of chlorine.
4. The cistern should then be topped up with water from the service pipe pipe and more chlorine added.
5. The cistern and pipes should then remain charged for at least three hours and then tested for residual chlorine; if none is found, the sterilisation will have to be carried out again.
6. Finally, the cistern and pipes should be thoroughly flushed out before any water is used for domestic purposes.

Exercises

1. Define
   (a) feed cistern,
   (b) storage cistern,
   (c) warning pipe,
   (d) supply pipe,
   (e) service pipe,
   (f) distributing pipe.
2. Sketch and describe the method of connecting the communicating pipe to the water main.
3. Describe the direct and indirect systems of cold water supply.
4. Make isometric diagrams of direct and indirect cold water systems for
   (a) a house with sink, bath, shower, basin and WC,
   (b) a three-storey office with three WCs, four wash basins, one sink and one drinking fountain on each floor.
5. Sketch the method of installing a cistern sited inside a roof space.
6. State the advantages and disadvantages of providing storage of cold water in buildings.
7. Explain how drinking water may become polluted by back siphonage and the methods used for its prevention.
8. Sketch the method used to interconnect two or more cisterns to provide a means of repairing, renewing or cleaning one of the cisterns whilst still maintaining a supply to the building.
9. Describe the precautions necessary in installing a cold water system to prevent damage to the system by the action of frost.
10. Sketch the method of installing a meter and state the types of building requiring a meter.
11. State the method used to estimate the amount of cold water storage in buildings.
12. Describe how a cold water system is sterilised.

Chapter 3 Boosted Cold Water Installations

Principles

For building constructed so that the water supply to them is above the level of the mains head, it will be necessary to provide boosting equipment. This problem arises in multi-storey buildings or buildings constructed on high ground.

If, for example, the pressure on the water main during the peak demand period is found to be 400 kPa, this pressure would supply water inside the building up to a height of about 40 m. In order to provide a good delivery of water at the highest fittings a residual head of 2 m above these fittings is usually required. In this example, therefore, the main would supply water up to a height of 40 − 2 = 38 m.

If the building is multi-storey with a vertical height from the main to the highest fittings of, say, 50 m, it will be necessary to provide boosting equipment to lift the water from 38 m to 50 m.

There are basically four systems in common use:

1. directly boosted system to cold water cisterns only,
2. directly boosted system to a header pipe,
3. indirect boosting from break cisterns at low level,
4. the auto-pneumatic system.

Directly Boosted System to Cisterns (Fig. 3.1)

Where the water regulations permit, pumps can be connected directly to the incoming main, thus enabling the pump head to be added to that of the mains. Control of the pump is effected by means of a float switch or electrode probes in the roof-level storage and drinking water header cistern. Fig. 3.2 shows a detail of the drinking water header cistern.

Directly Boosted System to Header Pipe (Fig. 3.3)

This system incorporates an enlarged header pipe to provide drinking water at high level between pump on-off cycles. The storage cisterns are provided with a float or electrode probes and the refilling of the header pipe is controlled by a pipeline switch (see Fig. 3.4). A time delay of approximately 2 min is fitted to the pipeline switch. As the header pipe is normally refilled in a shorter time than this, the excess water is discharged into the storage cisterns.

Centrifugal pumps, however, can operate against a closed valve for a short period of time and therefore no damage to the pumps is likely to occur if the ballvalves are closed during the refill cycle.

Indirect Boosting from Low-Level Cistern

Many water authorities require a break cistern to be installed between the main and the boosting unit. The cistern will serve as a

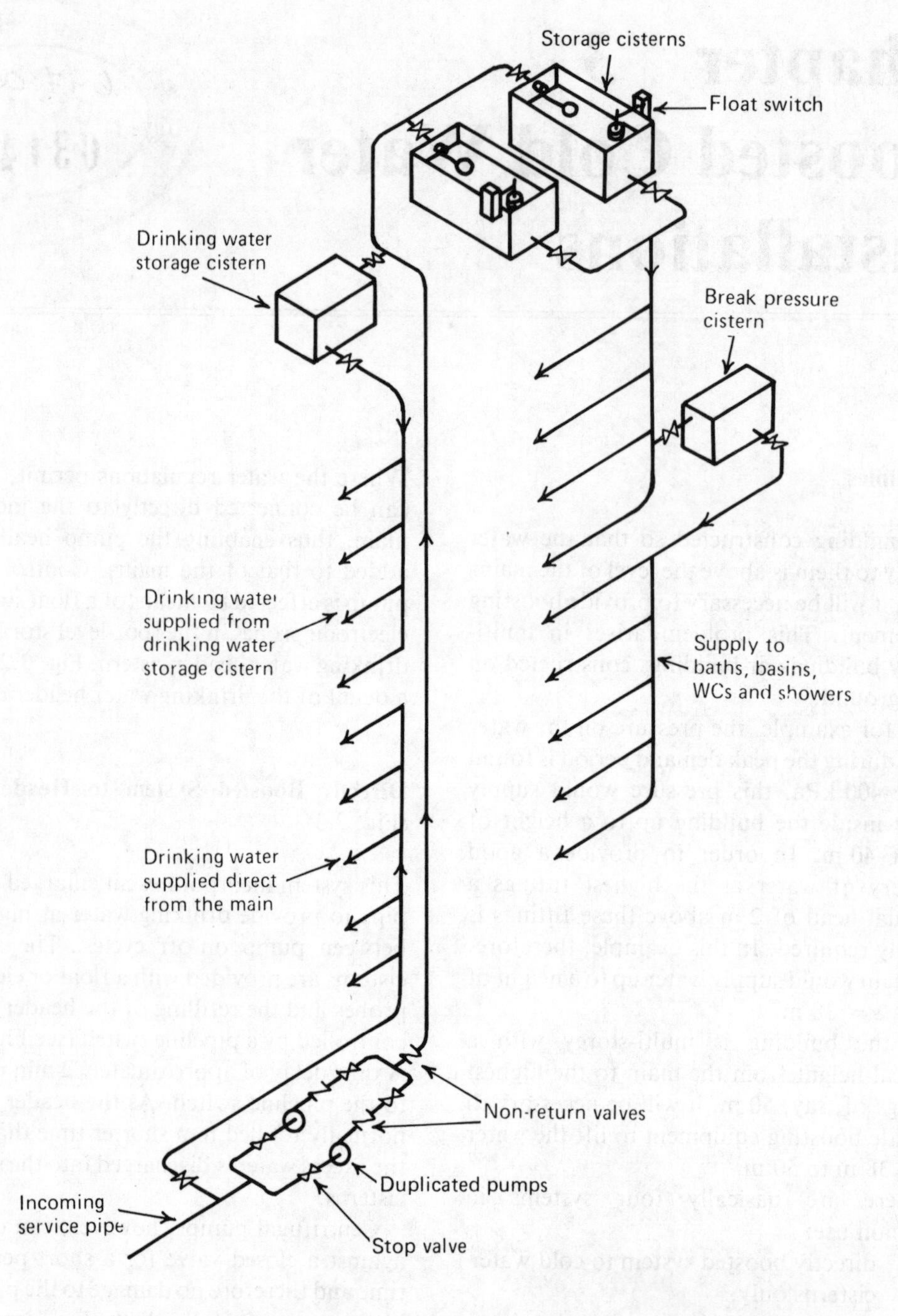

Fig. 3.1 Directly boosted to drinking water and storage cisterns.

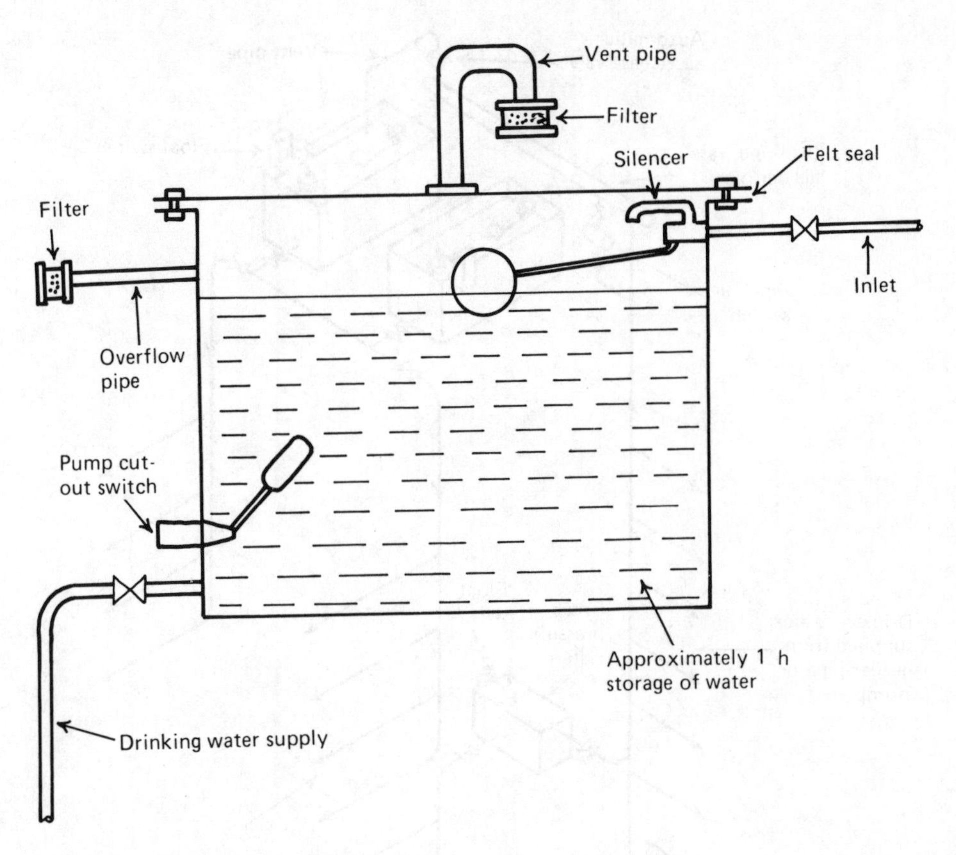

Fig. 3.2 Detail of drinking water header cistern.

boosting reservoir and prevent lowering of the pressure on the main. As there is no assistance from the mains pressure, the boosting equipment must be capable of overcoming the total static head of water plus the frictional resistances in the pipework.

The sizing of the break cistern must be considered carefully to prevent stagnation of water, which could occur because of oversizing. Where the whole of the storage is to be provided at low level, the local water authority must be consulted.

Float or electrode probe control switches must be provided to the break cistern to switch off the pumps when the water level drops to about 250 mm above the pump suction inlet. This precaution is necessary to prevent the pumps running dry during an interruption of supply. Fig. 3.5 shows details of the low-level break cistern.

Auto-pneumatic System (Fig. 3.6)

In an indirect system of cold water supply, a steel cylinder may be used as an alternative to a drinking water header cistern or pipe. The cylinder contains compressed air in the top which is pressured by the water pumped into the bottom. This cushion of air serves to force water up to the high-level drinking water points and storage cistern. When drinking water is drawn off through the high-level fittings, the water level in the cylinder falls. At a predetermined low level a pressure switch cuts in the pump and the cylinder is refilled up to a predetermined high level, when another pressure switch cuts out the pump.

Purpose of the Air Compressor. In time some of the air inside the cylinder is absorbed into the water and a gauge glass is usually fitted to give a visual indication of the water level. As the air becomes absorbed, a smaller quantity is available to provide the required reserve of pressure and the frequency of

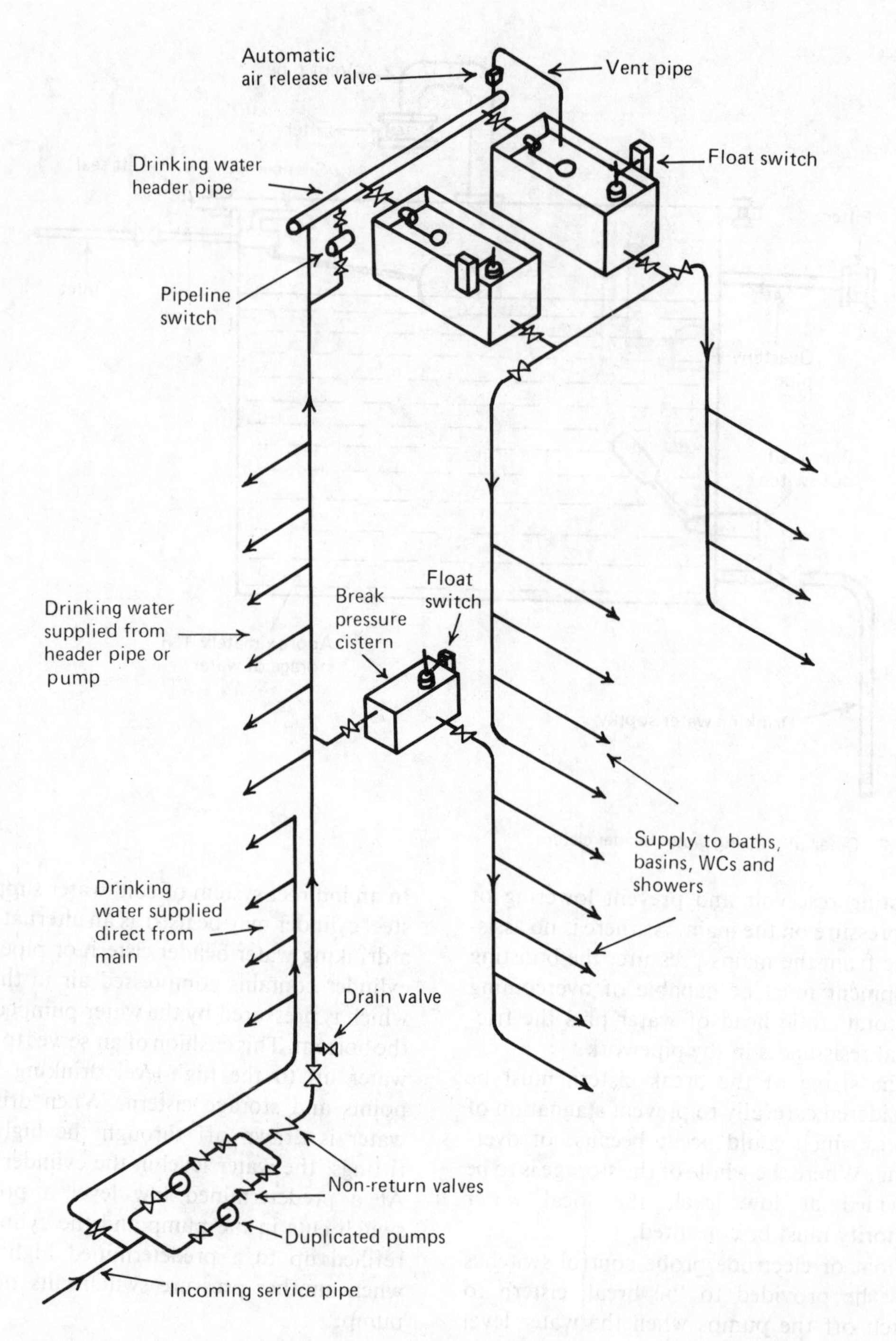

Fig. 3.3 Directly boosted to drinking water header pipe and storage cisterns.

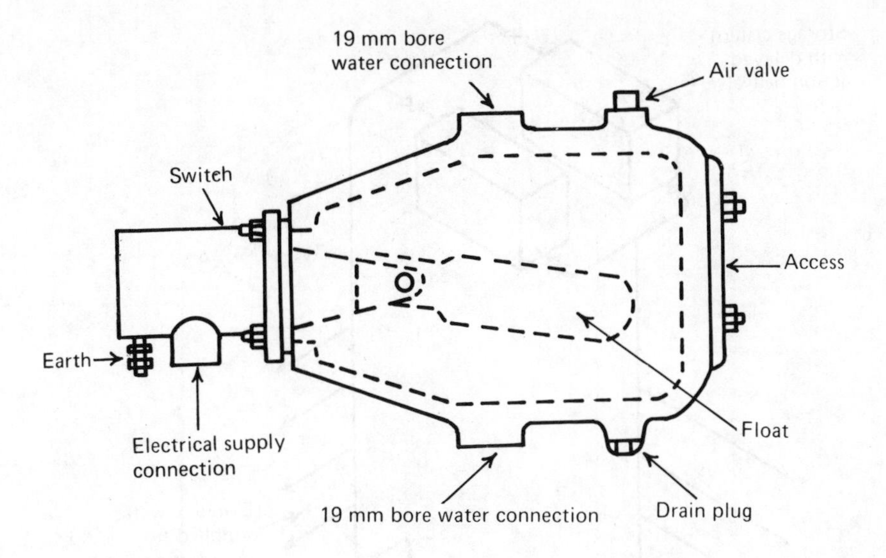

Fig. 3.4 Pipeline switch.

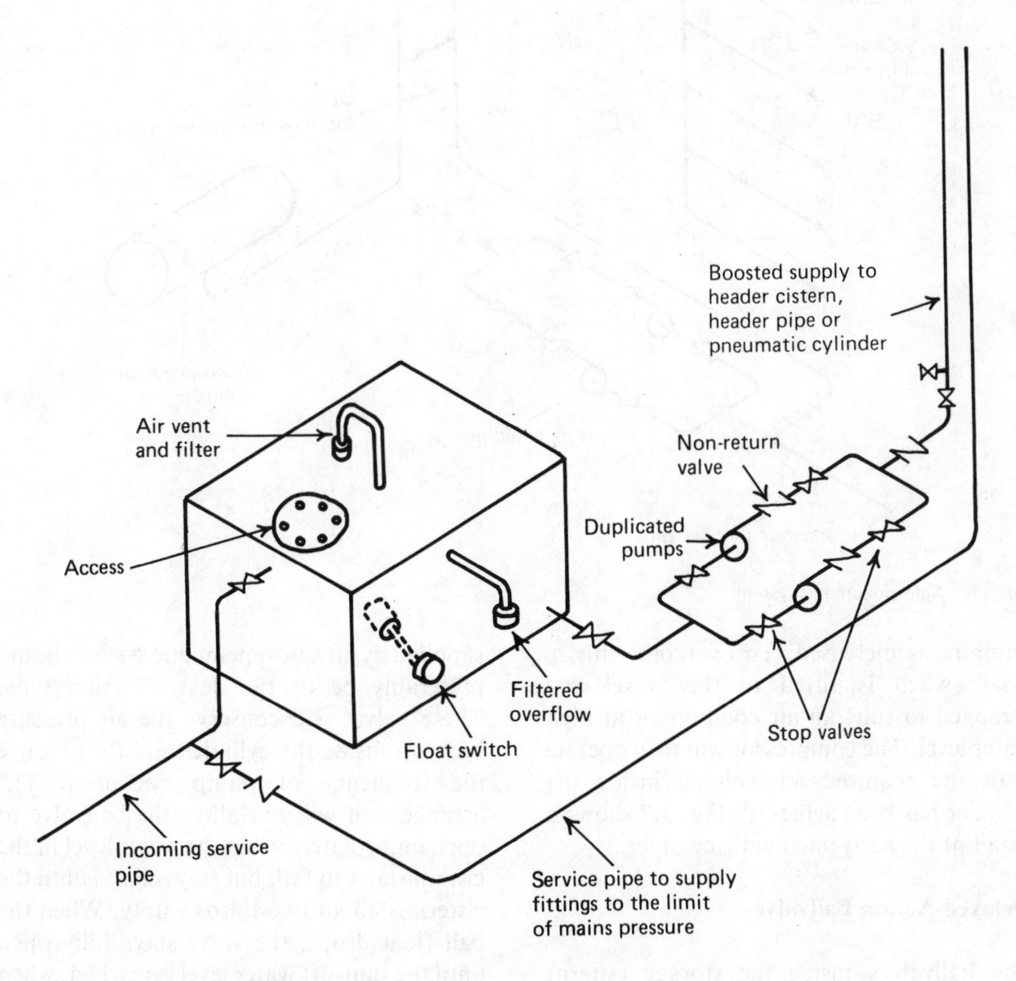

Fig. 3.5 Low-level break cistern.

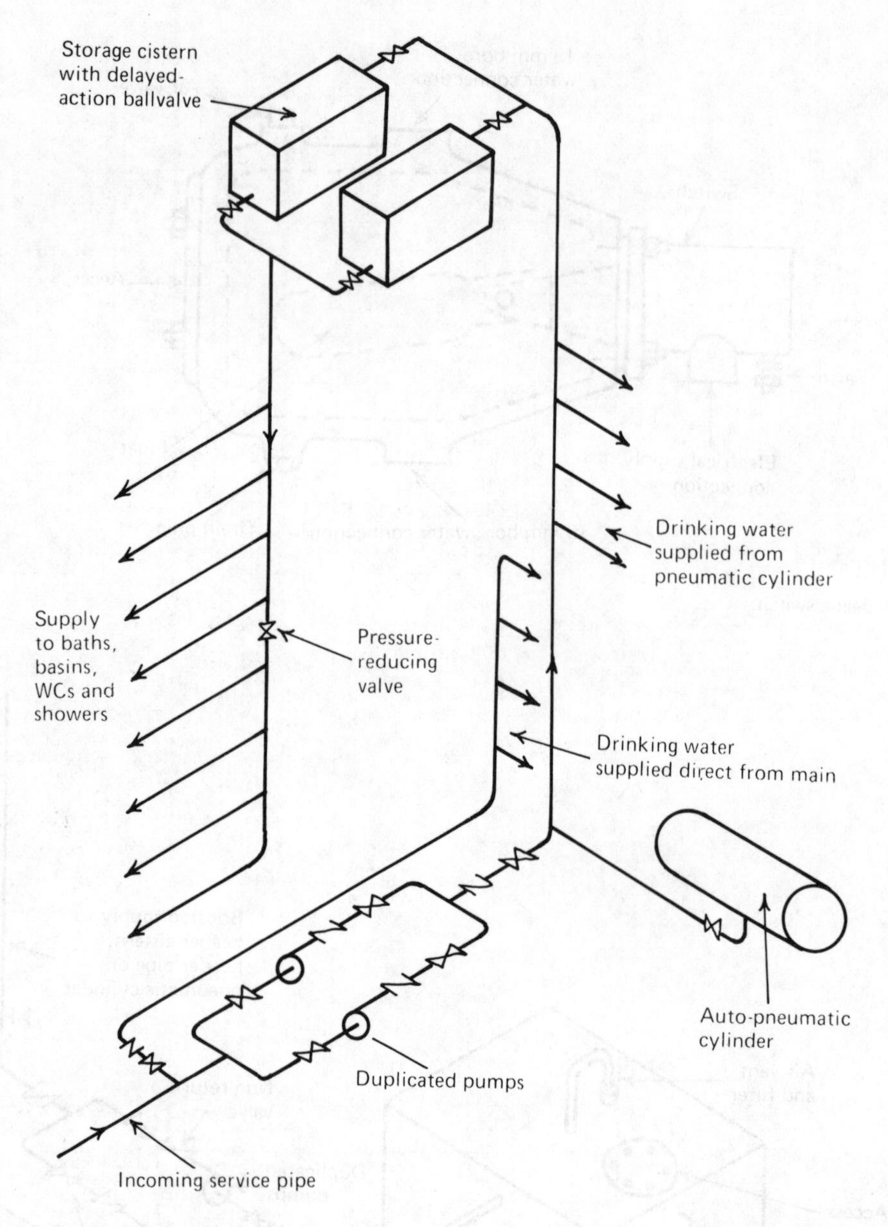

Fig. 3.6 Auto-pneumatic system.

pumping is increased. To overcome this, a float switch is fitted to the vessel and arranged to start an air compressor at high water level. The compressor will then operate until the required air volume inside the cylinder has been achieved. Fig. 3.7 shows a detail of the auto-pneumatic cylinder.

Delayed-Action Ballvalve

The ballvalves inside the storage cisterns supplied by an auto-pneumatic system should preferably be of the delayed-action type. These valves will conserve the air pressure built up inside the cylinder and thus reduce the frequency of pump operation. The arrangement will not allow the ballvalve to open immediately when the water level in the cistern starts to fall, but stays closed until the cistern is about two-thirds empty. When the ball float drops, the valve stays fully open until the shut-off water level is reached, when

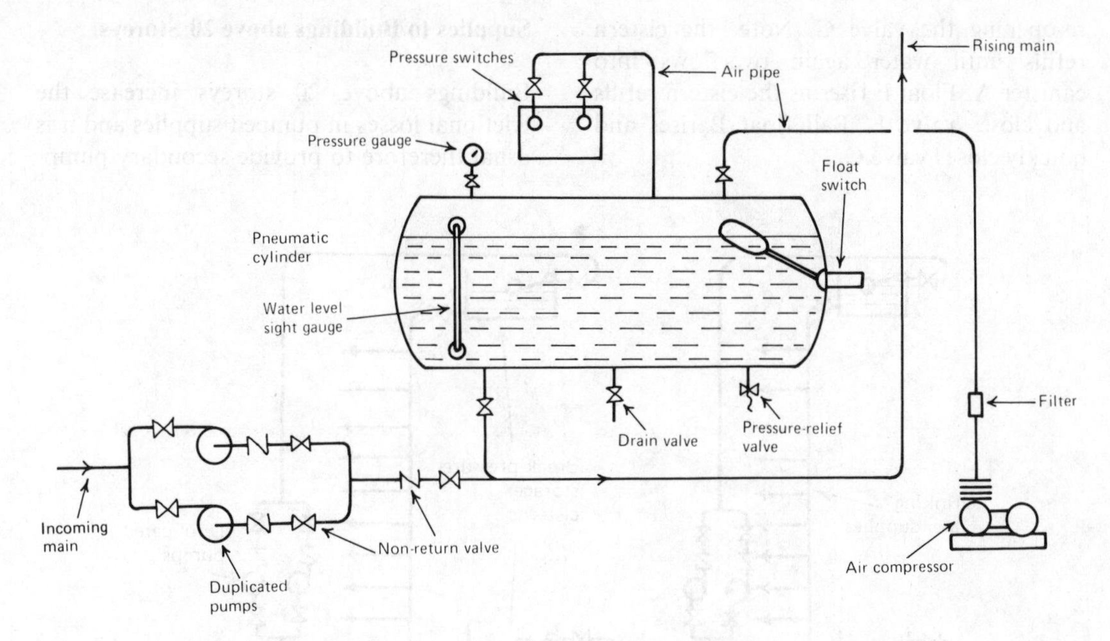

Fig. 3.7 Automatic pneumatic cylinder.

it closes quickly.

Fig. 3.8 shows the sequence of operation of the valve. Fig. 3.8a shows the cistern full, valve F closed and the water flowing over the rim into the ball float canister A. The ball float B is rising and closing the inlet valve C. In Fig. 3.8b, the cistern is being emptied and the ball float B is floating on the water, which remains in the canister A. Valves F and C remain closed. In Fig. 3.8c, water has been drawn off from the cistern to a predetermined level, which allows float E to be lowered. Valve F is now opened, which releases water from canister A and lowers ball float B thus

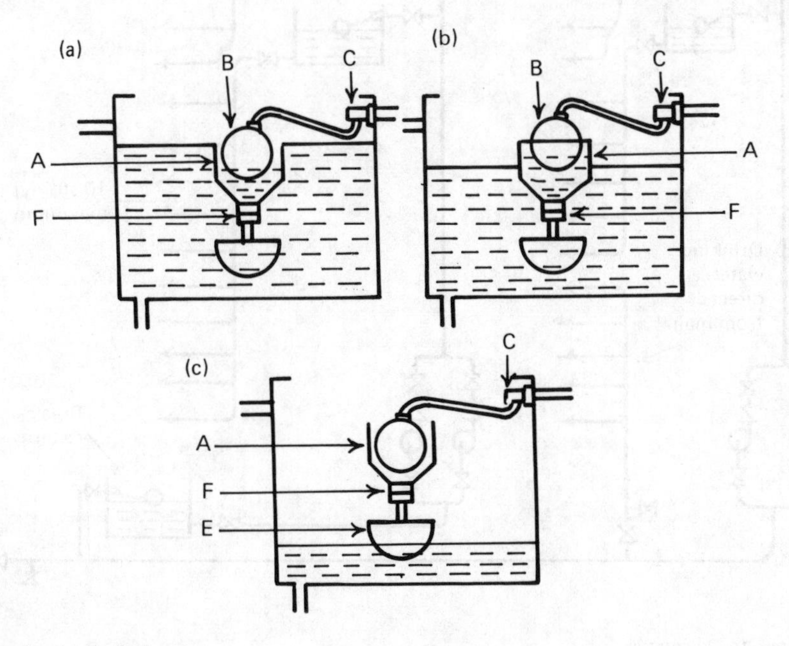

Fig. 3.8 Delayed-action ballvalve.

re-opening the valve C. Note: the cistern refills until water again overflows into canister A. Float E rises as the cistern refills and closes valve F. Ball float B rises and quickly closes valve C.

Supplies to Buildings above 20 Storeys

Buildings above 20 storeys increase the frictional losses in pumped supplies and it is usual therefore to provide secondary pump-

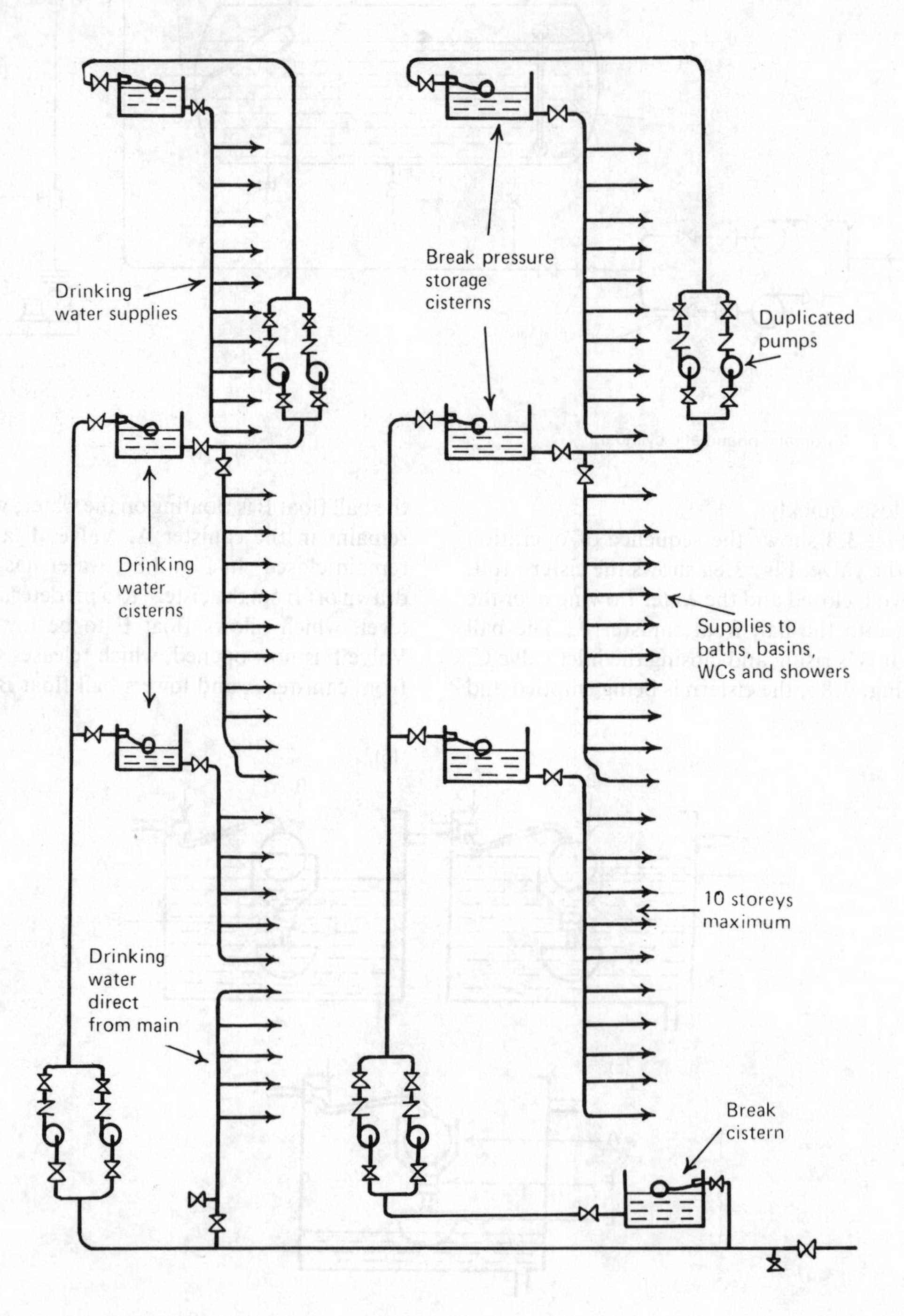

Fig. 3.9 System for 30 storeys.

ing equipment on the twentieth floor. Fig. 3.9 shows a system for 30 storeys.

Distribution from Storage Cisterns

In order to avoid excessive pressures in the pipework, the maximum head of water in the system must be limited to 30 m. The floors of a multi-storey building must therefore be zoned by means of a break-pressure cistern or pressure-reducing valve. Figs. 3.1, 3.3 and 3.9 show the method of zoning by break-pressure cisterns and Fig. 3.6 shows zoning by use of a pressure-reducing valve.

Exercises

1. Sketch sections through the following and describe their operation:
   (a) break cistern,
   (b) delayed action ballvalve,
   (c) pipeline switch.
2. Sketch and describe the operation of a cold water installation for a 16-storey building when the water is to be pumped:
   (a) directly from the main,
   (b) indirectly from the main.
3. Sketch a section through a pneumatic cylinder and describe its operation.
4. State the purpose of the following and explain their operation:
   (a) break cistern,
   (b) break-pressure cistern,
   (c) float switch,
   (d) automatic air valve.
5. Sketch and describe a cold water installation for a 30-storey building.
6. Sketch a method of interconnecting two pumps so that one of the pumps may be used for stand-by purposes.

Chapter 4
Fire Control Systems

Hose Reels

These may be used by the occupants of the building as a first aid and it is often possible to extinguish a fire by a jet of water before the fire brigade arrives. It is, however, possible to extinguish a fire by means of a suitable portable extinguisher and for this reason these should not be dispensed with when hose reels are installed.

Positioning. As hose reels are intended for use by the occupants of the building, they should be sited in positions that will be accessible without exposing the user to danger from the fire. For this reason they are usually fixed along escape routes or close to the fire exits, so that personnel escaping from a fire will pass them on their way to safety and may use them without having their means of escape cut off.

In office blocks, especially the multi-storey type, the hose reels must be fitted inside the actual office accommodation, which means that they are usually fitted adjacent to the fire exit doors. This enables the hose reels to be used without opening the smoke stop doors of the escape lobby and therefore prevents the lobby being filled with smoke.

In industrial buildings it is not always possible to fix hose reels only near to the exit doors, owing to the fact that the width of the building may prevent the hose nozzle reaching a fire in the centre. In these buildings it is also necessary to fix additional hose reels in the centre of the building, usually on the columns.

Design Considerations. Hose reel installations should be designed so that no part of the floor is more than 6 m from the hose nozzle when the hose is fully extended. The water supply must be able to provide a discharge of not less than 0.4 l/s through the nozzle and also designed to allow not less than three hose reels to be used simultaneously at a flow rate of 1.2 l/s. A water pressure of 200 kPa is required at the nozzle and with this pressure the jet will have a horizontal distance of 8 m and a height of about 5 m.

Pipe Size. The Fire Offices Committee Rules require that pipes supplying hose reels should be not less than 50 mm in diameter and that the connection to the hose reel should not be less than the nominal bore of the hose.

It is usual to use a 50 mm diameter pipe for buildings up to 15 m in height and a 64 mm diameter pipe for buildings above 15 m in height. In some areas, the minimum diameter of the pipe connected to each hose reel should not be less than 25 mm in diameter.

Water Supply. If the supply authority's main can supply a minimum pressure at the highest reel of 200 kPa and also provide sufficient discharge of water, the hose reels may be supplied directly from the main (see Fig. 4.1).

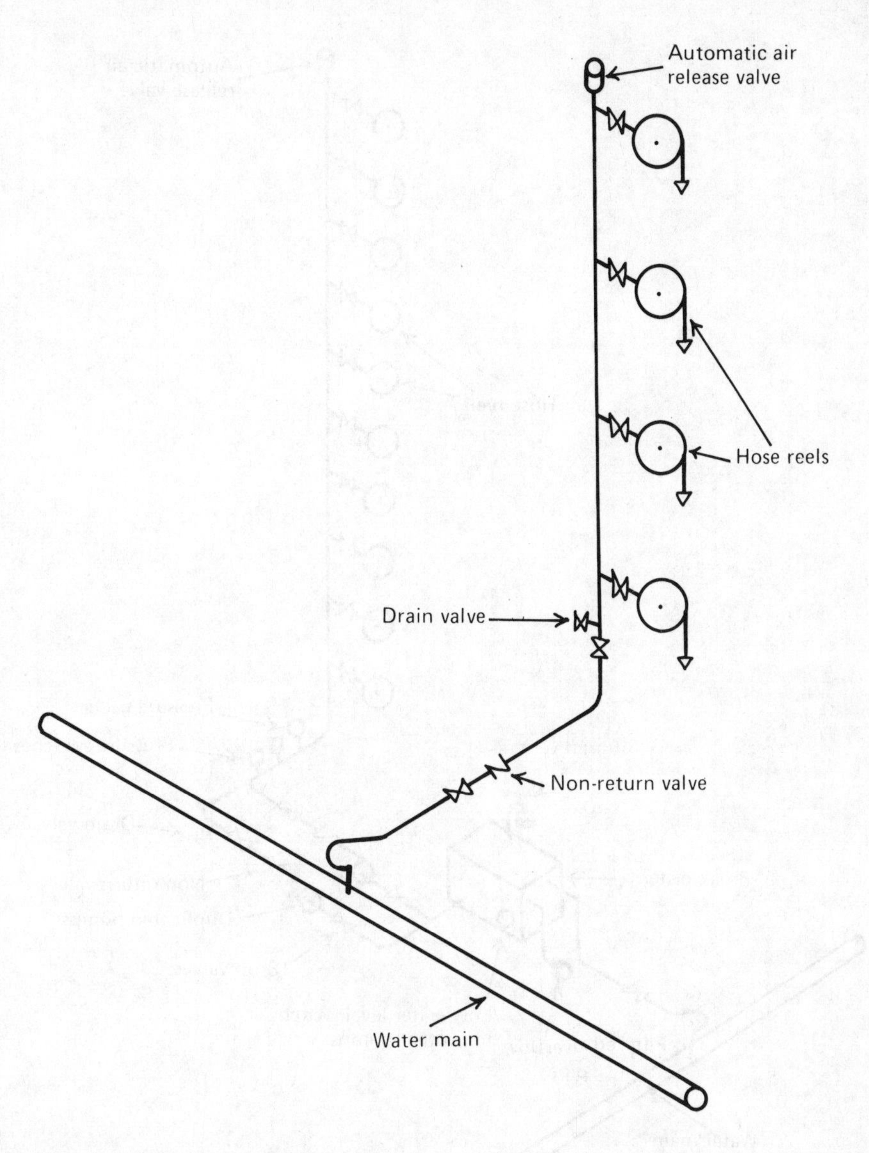

Fig. 4.1 Hose reel installation without pumping equipment.

If, however, the supply authority's main will not satisfy the required conditions, automatic pumping equipment will have to be installed. Some authorities will allow pumping equipment to be connected directly to the water main, provided that a reasonable flow would take place under mains pressure at the highest hose reel without pumping.

If the water supply authority requires a break cistern, this should hold a minimum of 1.6 m^3 and duplicated pumps should be installed that will provide a minimum discharge of 2.3 l/s. In large buildings a standby pump operated by a diesel engine may be required.

Fig. 4.2 shows a hose reel installation with pumping equipment. When a reel is used, the drop in water pressure allows one of the pressure switches to switch on the duty pump. As an alternative to a pressure switch, a flow switch inserted in the main on the delivery pipe from the pump may be used. The switch is capable of sensing a flow of 0.1 l/s and will keep the duty pump running until the hose reel is shut off.

Fig. 4.3 shows the installation of a fixed-

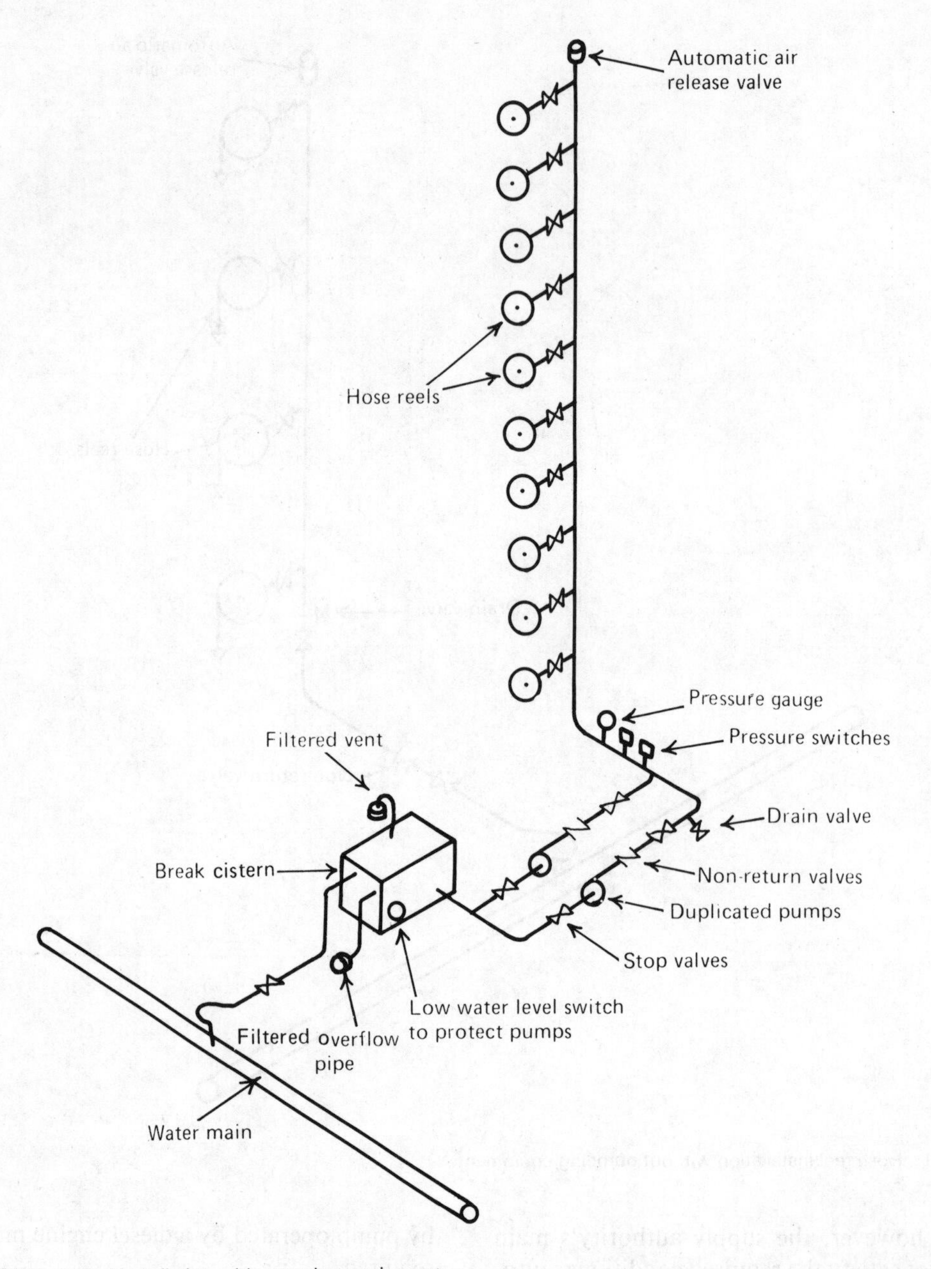

Fig. 4.2 Hose reel installation with pumping equipment.

type hose reel: a swinging-type hose reel may also be installed inside a corridor recess.

Dry Riser

A dry riser consists of an empty pipe rising vertically inside a building with landing valves connected to it on each floor and at roof level. An inlet is provided at ground level to allow the fire brigade to pump water into the riser from the nearest hydrant.

Dry risers are provided solely for the use of the fire brigade and are therefore not installed for first aid use by the building occupants in the same way as a hose reel. The dry riser is really an extension of the fire brigade hose

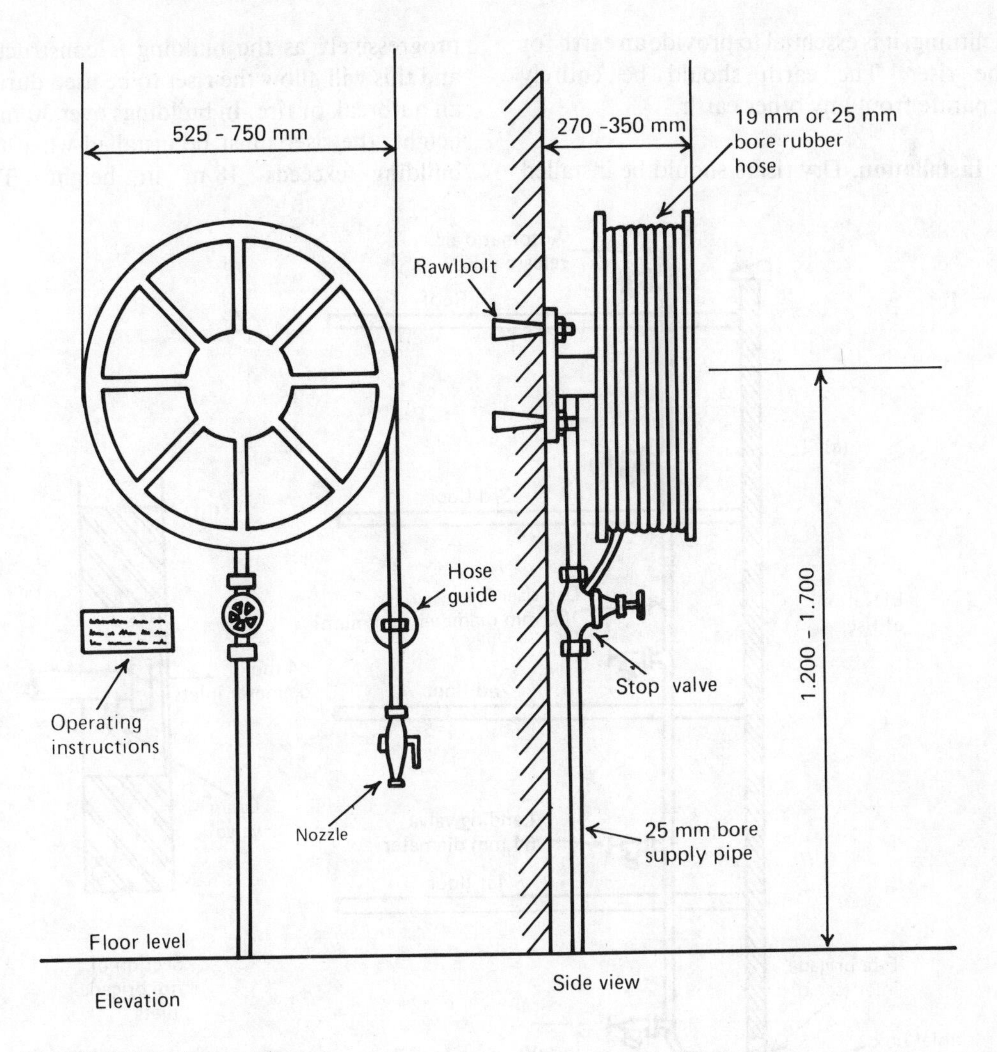

Fig. 4.3 Detail of fixed hose reel.

and avoids the necessity of running long lengths of canvas hose up the staircase of a building from ground level.

Positioning. Dry risers are usually positioned in a ventilated lobby approach to the staircase and this enables the fire brigade to connect their hose pipes to a landing valve in a smoke-free area.

Diameter of Riser. In buildings up to 45 m in height and where there is one 64 mm landing valve on each floor, the internal diameter of the riser should be 100 mm. In buildings between 45 m and 60 m in height, the internal diameter of the riser should be 150 mm. A 150 mm riser is also required for any building having two 64 mm landing valves on each floor. It is not permissible to install a dry riser in a building above 60 m in height and a wet riser is required for these buildings.

Number of Risers. Dry rising mains should be disposed so that no part of the floor is more than 61 m from a landing valve, the distance measured along a route suitable for a hose line including any distance up and down a stairway. Outlets should be provided for every 930 m^2 of floor area from the ground level to the roof.

Earthing. In order to prevent the risk of electric shock and damage to the riser by lightning, it is essential to provide an earth for the riser. The earth should be entirely separate from any other earth.

Installation. Dry risers should be installed progressively as the building is constructed and this will allow the riser to be used during an outbreak of fire. In buildings over 30 m in height, the riser must be installed when the building exceeds 18 m in height. The

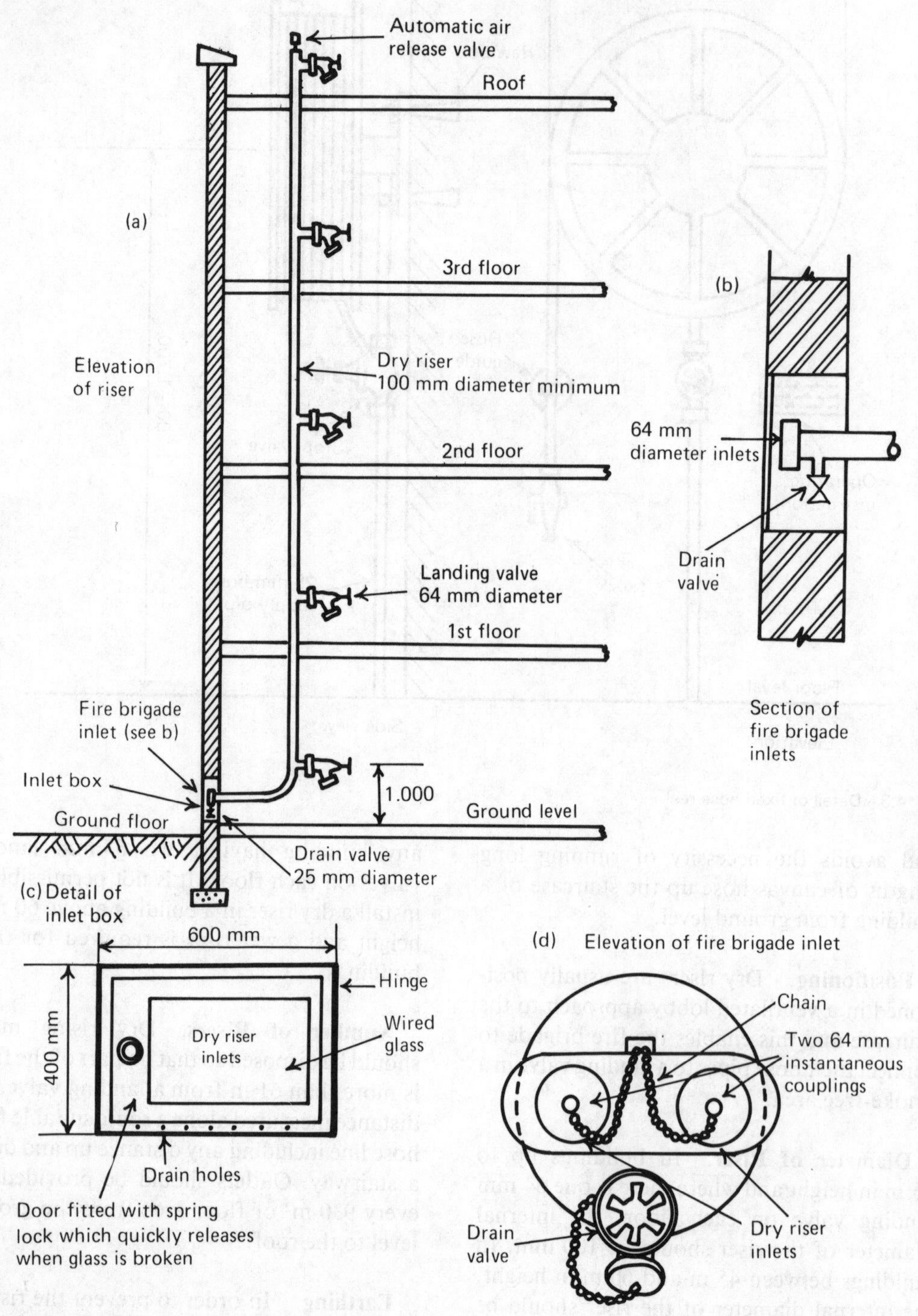

Fig. 4.4 Details of dry riser.

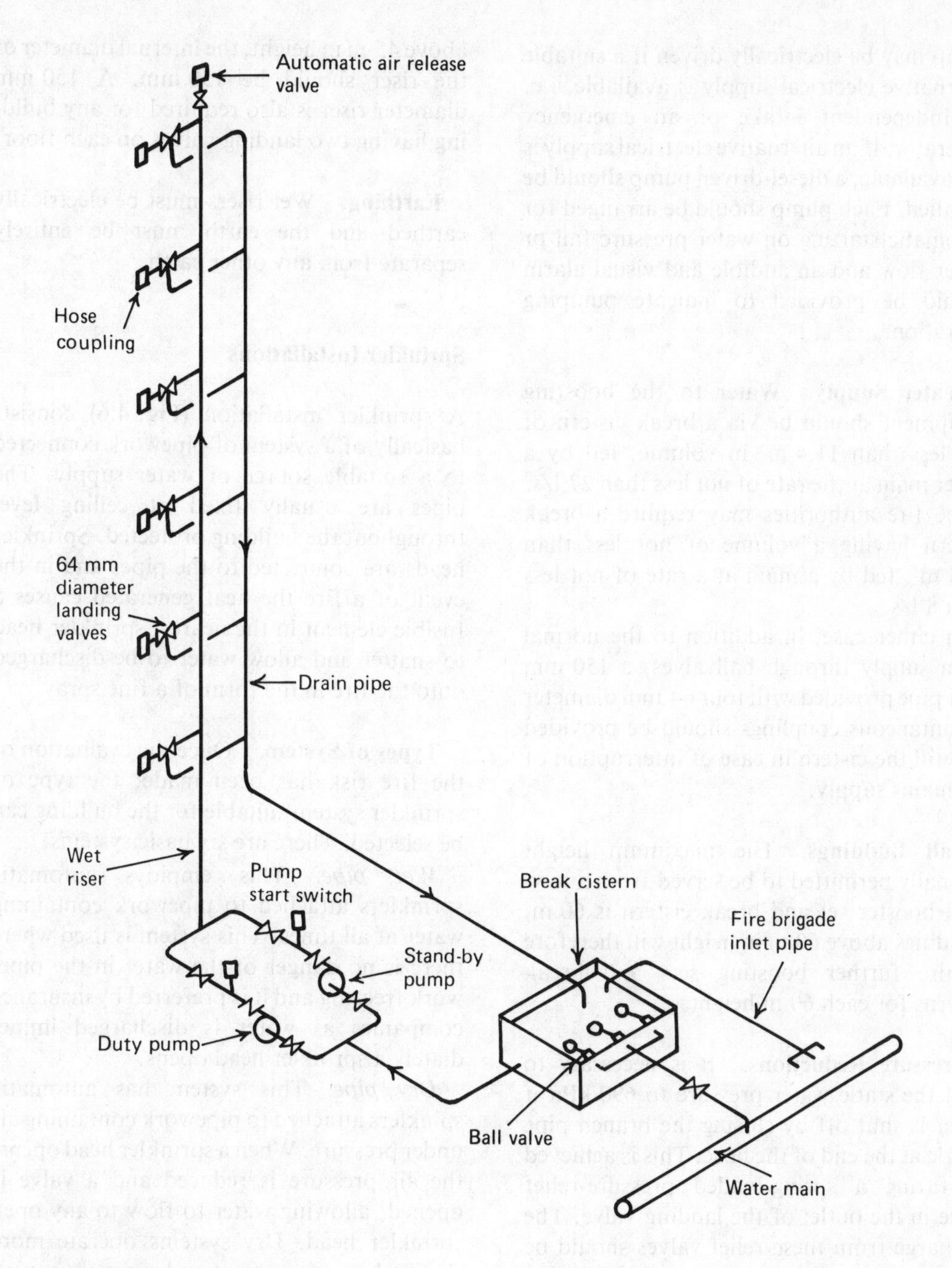

Fig. 4.5 Installation of a wet riser.

completed installation should be tested and approved by the local fire and water authorities. Fig. 4.4 shows the installation of a dry riser.

Wet Riser (Fig. 4.5)

These are permanently connected to a water supply capable of providing a running pressure of not less than 410 kPa at the top outlet. The maximum running pressure permitted with one outlet in operation is 520 kPa. To maintain the above running pressure it is necessary to employ boosting equipment capable of delivering 23 l/s. Duplicate pumping equipment is required and the stand-by pump may be electrically driven if a suitable alternative electrical supply is available, i.e. an independent intake or an emergency generator. If an alternative electrical supply is not available, a diesel-driven pump should be installed. Each pump should be arranged for automatic starting on water pressure fall or water flow and an audible and visual alarm should be provided to indicate pumping operation.

Water Supply. Water to the boosting equipment should be via a break cistern of not less than 11.4 m^3 in volume, fed by a water main at the rate of not less than 27 l/s. Some fire authorities may require a break cistern having a volume of not less than 45.5 m^3, fed by a main at a rate of not less than 8 l/s.

In either case, in addition to the normal main supply through ballvalves, a 150 mm inlet pipe provided with four 64 mm diameter instantaneous couplings should be provided to refill the cistern in case of interruption of the mains supply.

Tall Buildings. The maximum height normally permitted to be served from a low-level booster set and break cistern is 60 m. Buildings above 60 m in height will therefore require further boosting sets and break cisterns for each 60 m height.

Pressure Reductions. It is necessary to limit the static water pressure to 690 kPa if water is shut off by closing the branch pipe nozzle at the end of the hose. This is achieved by fitting a spring-loaded pressure-relief valve in the outlet of the landing valve. The discharge from these relief valves should be piped via a 50 mm diameter connection to a 76 mm or 100 mm diameter drain pipe, taken vertically down the building along side the wet riser and discharged over the break cistern.

Diameter of Riser. In buildings up to 45 m in height and where there is one landing valve on each floor, the internal diameter of the riser should be 100 mm. In buildings above 45 m in height, the internal diameter of the riser should be 150 mm. A 150 mm diameter riser is also required for any building having two landing valves on each floor.

Earthing. Wet risers must be electrically earthed and the earth must be entirely separate from any other earth.

Sprinkler Installations

A sprinkler installation (Fig. 4.6) consists basically of a system of pipework connected to a suitable source of water supply. The pipes are usually fixed at ceiling level throughout the building protected. Sprinkler heads are connected to the pipes and in the event of a fire the heat generated causes a fusible element in the nearest sprinkler head to shatter and allow water to be discharged onto the fire in the form of a fine spray.

Types of System. Once the evaluation of the fire risk has been made, the type of sprinkler system suitable for the building can be selected. There are six basic systems.

Wet pipe. This employs automatic sprinklers attached to pipework containing water at all times. This system is used where there is no danger of the water in the pipework freezing and it is preferred by insurance companies as water is discharged immediately a sprinkler head opens.

Dry pipe. This system has automatic spinklers attached to pipework containing air under pressure. When a sprinkler head opens, the air pressure is reduced and a valve is opened, allowing water to flow to any open sprinkler head. Dry systems operate more slowly than wet systems and are more expensive to install and maintain. For these reasons they are normally installed only where there would be a risk of water in the pipework freezing.

Alternate wet and dry. This system is used in unheated buildings and operates as a wet system during the summer months. When winter approaches, the pipework is drained of water and filled with compressed air so

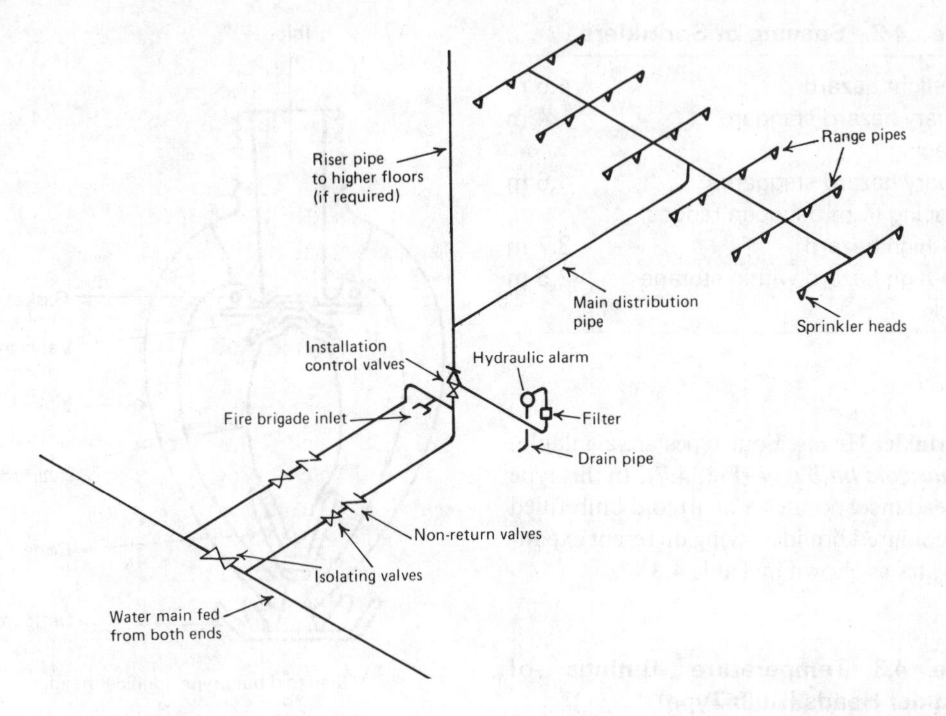

Fig. 4.6 Typical wet-pipe sprinkler installation.

that it operates as a dry system during the winter months.

Pre-action systems. These are designed primarily to counteract the operational delay in the dry pipe system and also to eliminate the risk of water damage resulting from accidental damage to sprinkler heads or piping. In the pre-action system, the water supply valve is actuated independently of the opening of the sprinkler heads. The water supply valve is instead opened by the operation of an automatic fire detection system and not by the opening of the sprinkler heads.

Deluge system. The purpose of the deluge system is to deliver the maximum amount of water in the minimum time. The system allows water to cover an entire fire area by admitting water to sprinkler heads or spray nozzles which are open at all times. By using automatic fire detection devices, it is possible to apply water to a fire more quickly than with systems that depend on the opening of sprinkler heads. The system is suitable for extra hazard occupancies in which flammable liquids are stored or handled and where there is a risk that fire may flash ahead of the operation of conventional sprinklers.

Fire-cycle system. In its initial operation, this system is the same as the pre-action system. It has, however, the additional ability to cycle on and off while controlling the fire and to shut itself off when the fire has been extinguished. The system therefore drastically reduces water damage and the on-and-off operation also permits sprinkler heads to be replaced without the necessity of shutting off the main supply valve.

Sprinkler Coverage. The maximum floor area covered by a single sprinkler depends on the fire hazard classification of the building (see Table 4.1). The maximum spacing of sprinklers is as shown in Table 4.2.

Table 4.1 Fire Hazard Classification

Extra-light hazard	21 m²
Ordinary hazard	12 m²
Extra-high hazard	9 m²
Extra-high hazard within storage racks	7.5 m²

Table 4.2 Spacing of Sprinklers

Extra-light hazard	4.6 m
Ordinary hazard standard spacing	4 m
Ordinary hazard staggered spacing (4 m between ranges)	4.6 m
Extra-high hazard	3.7 m
Extra-high hazard within storage racks	2.5 m

Sprinkler Heads. Four types are available.

Quartzoid bulb type (Fig. 4.7). In this type the head incorporates a quartzoid bulb filled with coloured liquids having different expansion rates as shown in Table 4.3.

Table 4.3 Temperature Ratings of Sprinkler Heads (Bulb Type)

Bulb rating (°C)	Colour of bulb liquid
57	Orange
68	Red
79	Yellow
93	Green
141	Blue
182	Mauve
227/288	Black

Side-wall type. This is the same as the quartzoid bulb type but is provided with a deflector at its base so that the spray of water is thrown to one side. They are designed for use at the side of corridors or rooms.

Soldered strut type. The soldered strut consists of three bronze plates joined together by a special low-melting-point solder. These plates hold a glass valve in position against an inlet orifice in a flexible diaphragm and seal the water outlet. When the solder is melted by heat from a fire, the plates fall apart releasing the glass valve and allowing water to spray over the fire.

Duraspeed type. This is an improved version of the simple soldered strut type. The

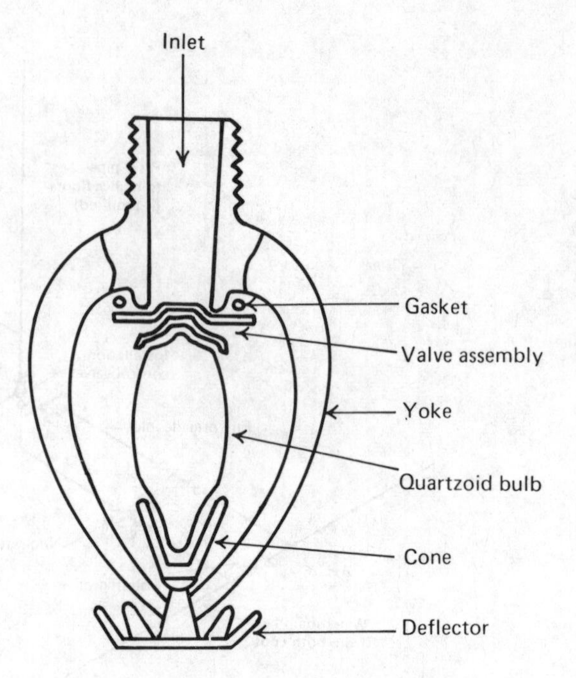

Fig. 4.7 Quartzoid bulb-type sprinkler head.

solder joining together the struts is almost enclosed inside a special element and this reduces the risk of premature operation of the head due to corrosion of the solder. A protective film is applied to the soldered element as a further precaution against atmospheric corrosion.

Sprinkler heads are manufactured with nominal orifice diameters to suit the respective fire hazard classification as shown in Table 4.4.

Table 4.4 Nominal Orifice Diameters of Sprinkler Heads

N.B. Sprinkler heads must never be painted.

Nominal diameter (mm)	Hazard classification
10	Extra light
15	Ordinary
20	Extra high

Water Supply for Fire-Control Systems

Water supply from the main, either direct or via a break cistern, and from a high-level storage cistern have been dealt with in the previous descriptions. Other water supply sources that may be suitable are:

1. supply from main and pneumatic cylinder as described in Chapter 3;
2. elevated private reservoir;
3. river or canal.

For the highest insurance rebate, two distinct supplies of water are usually required, i.e. a break cistern supplied from the main and an elevated private reservoir.

Exercises

1. Sketch and describe the operation of a hose reel installation for buildings where
   (a) the water mains pressure is sufficient,
   (b) the water mains pressure is insufficient.
2. State the hydraulic requirements for a hose reel installation.
3. Sketch a hose reel installation supplied from a high-level break cistern.
4. State the positions in a building where hose reels should be sited.
5. Sketch and describe the operation of a dry riser for a 10-storey building.
6. State the purpose of a dry riser.
7. State the diameters of dry risers and the diameters of the fire brigade inlets.
8. State the purpose of a wet riser.
9. Sketch and describe the operation of a wet riser for a 25-storey building.
10. State the hydraulic requirements of a wet riser.
11. Describe the following sprinkler installations:
    (a) wet pipe,
    (b) dry pipe,
    (c) alternate wet and dry pipe,
    (d) pre-action,
    (e) deluge,
    (f) fire cycle.
12. Sketch a section through the following types of sprinkler head and describe their operation:
    (a) quartzoid bulb type,
    (b) soldered strut type.
13. State the maximum spacings of sprinkler heads.
14. State the various water supply sources that may be used for fire-control systems and the type of supply that will allow the highest insurance rebate.

Chapter 5
Natural Gas, Service and Installation Pipework

Formation of Natural Gas

Throughout the world there occur in certain rock formations accumulations of a complex mixture of hydrocarbons to which the name 'petroleum' has been given and, in its natural state, petroleum can exist in solid, liquid or gaseous form, though all are of the same origin.

Solid or semi-solid petroleum is composed of heavier hydrocarbon compounds. It is usually found at the surface of the earth where the lighter hydrocarbons from liquid petroleum near the surface have escaped to the atmosphere. This type of petroleum is called asphalt, tar, pitch or bitumen, and is used by builders for weathering and waterproofing. A good example of this is the asphalt lake in Trinidad.

Natural liquid petroleum is known as 'crude oil' and consists of liquid hydrocarbons, which occur below the surface of the earth, although it sometimes escapes through fissures to the surface. Crude oil is the raw material for oil refineries where fractional distillation separates off petrol, gas oil, diesel oil and fuel oil, leaving pitch or bitumen as the residue.

Characteristics of Natural Gas

Gaseous petroleum is called 'natural gas' and consists of lighter hydrocarbons, which remain gaseous under the conditions of their occurrence, or become gaseous when they reach the surface. The gas consists of methane, ethane, propane and butane, and usually consists of 90% or more methane. The gas usually occurs below the surface of the earth, although like crude oil it sometimes escapes to the surface and evaporates to the atmosphere.

Commercial deposits of petroleum can exist only if geological conditions are favourable in the first place for their formation. The world's accumulation originates in the deposition of marine organisms, after death, on the sea bed. These organisms, which include plankton, bacteria and simple plants, accumulate with large quantities of inorganic particles deposited by rivers or through chemical precipitation.

A second possible source of natural gas is deep coal measures, where petroleum can be formed under the action of heat and pressure. Once the hydrocarbons have been formed, commercial quantities can accumulate only if there is a suitable trap in which they can collect. As soon as the reactions produce light hydrocarbons, these begin to move upwards, through the salt water that fills the minute pore space in the rock, and into adjacent formations. If, at this stage, a well were to be sunk into the stratum, production would consist of salt water with traces of oil and gas. If, however, the light hydrocarbons eventually meet an impermeable layer of rock salt, or shale, then an oil or gas field will be formed (see Fig. 5.1).

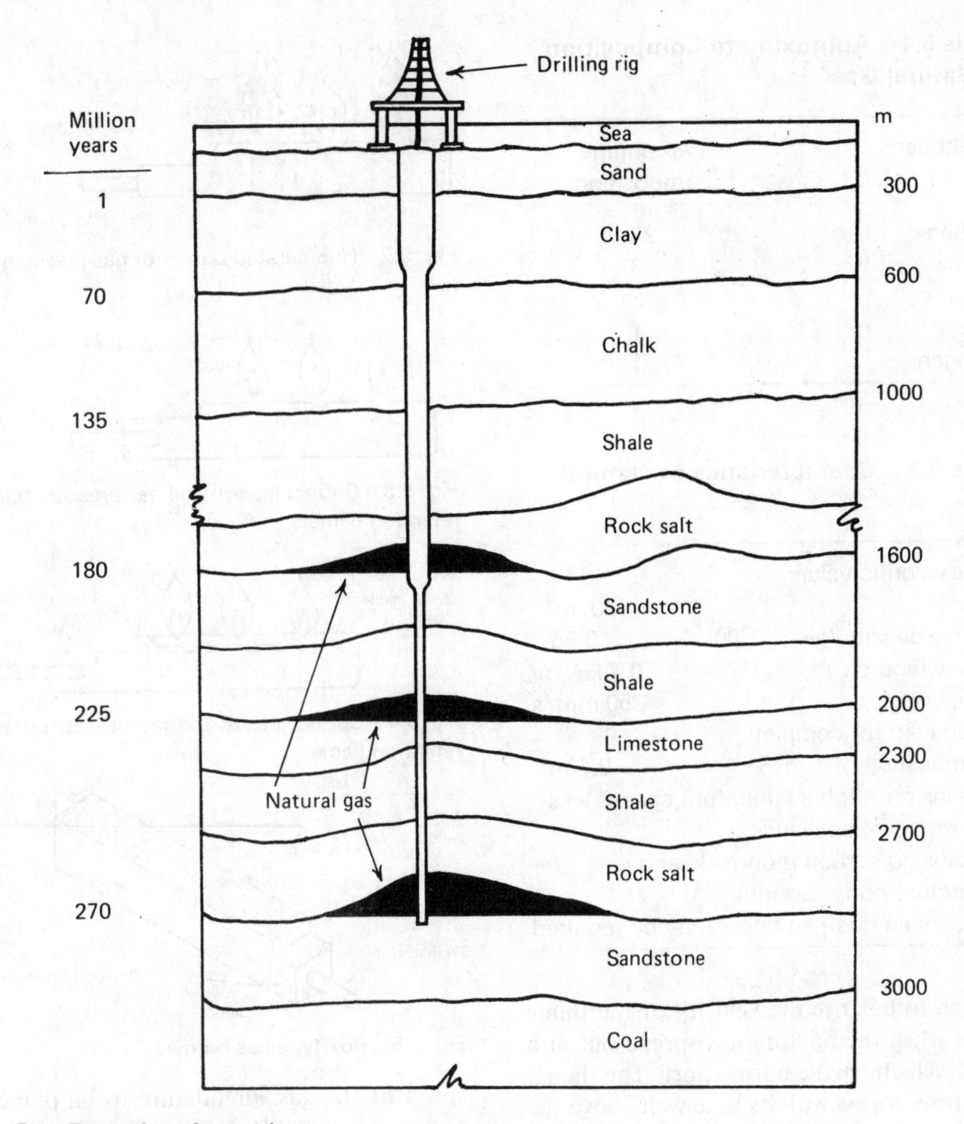

Fig. 5.1 Formation of natural gas.

These structural traps can also be formed by geological faults, which have the effect of sealing off the reservoir rock, or by salt domes formed by the upward intrusion of salt.

Table 5.1 gives the approximate composition of natural gas and Table 5.2 its characteristics.

Gas Burner

Since natural gas has a very slow burning velocity, there is a tendency for the flame to lift off the burner. When light is applied to a mixture of gas and air, the flame formed at the point of ignition spreads through the mass of gas and air at a definite maximum speed, which is known as the burning velocity of the gas.

A burner designed for a specific gas arranges for the speed of travel of the flame to be about the velocity of the mixture of gas and air issuing from the burner, when the flame has reached an area near the port of the burner. This equilibrium between the two velocities propagates a stable flame continuously. If, however, the burning velocity is very low, the velocity of the air and gas mixture from the burner will only have slowed up

Table 5.1 Approximate Composition of Natural Gas

Constituent	Percentage composition
Methane	93
Ethane	3
Propane	2
Butane	1
Nitrogen	1

Table 5.2 Characteristics of Natural Gas

Gross calorific value	37 MJ/m^3
Relative density (air = 1.00)	0.55
Density (approx.)	0.7 kg/m^3
Flame velocity	350 mm/s
Air required for complete combustion of 1 m^3	9.5 m^3
Contains no sulphur, therefore causes less corrosion of appliances	
Contains no carbon monoxide and is therefore nonpoisonous	
May contain dust, so filters may be required	

enough to balance the velocity of the flame travel when the mixture has spread out at a point well above the burner port. The flame, therefore, forms with its base well above the port, i.e. the flame lifts off.

Natural gas requires an aerated burner: otherwise the gas will not burn completely and the flame will be floppy and rather smoky. The gas pressure and injector jet have therefore to be such that the correct amount of air for complete combustion is achieved. Figs. 5.2, 5.3 and 5.4 show the results of incorrect and correct flame conditions as described.

The tendency of natural gas flames to lift off the burner may also be prevented by increasing the number of ports and therefore the total flame port area. This type is often called a box-type burner and consists of a perforated sheet-metal burner which allows the speed of the gas/air mixture to be reduced and stabilises the flame (see Fig. 5.5).

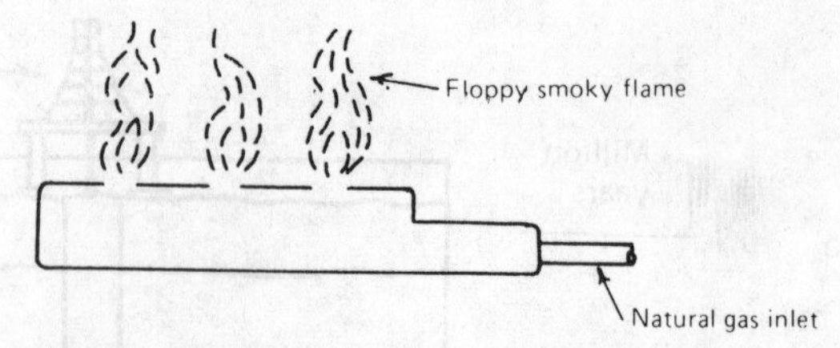

Fig. 5.2 Non-aerated burner or gas pressure too low.

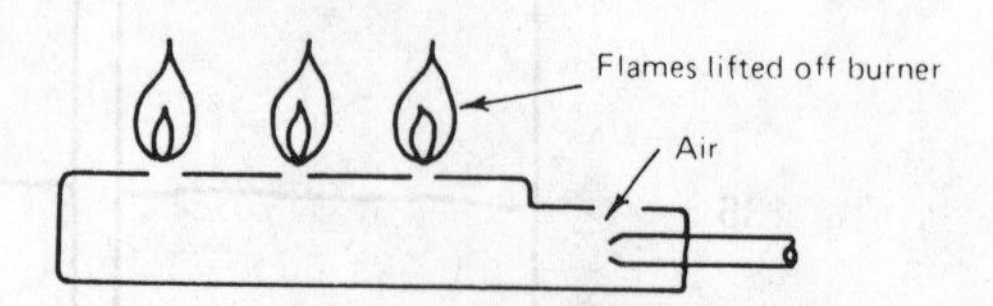

Fig. 5.3 Correct jet size and gas pressure but no retention flame.

Fig. 5.4 Correct jet size and gas pressure but with retention flame.

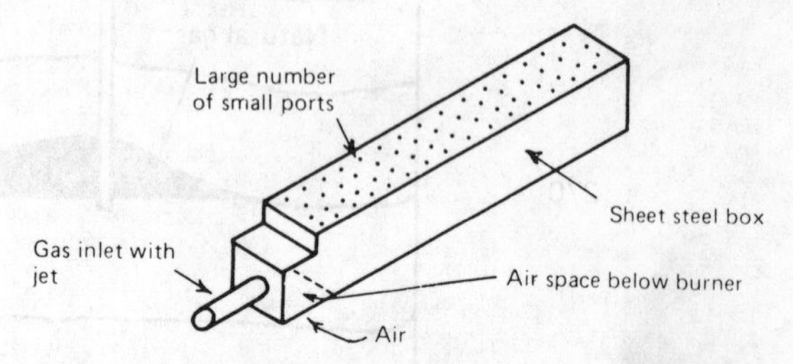

Fig. 5.5 Box-type gas burner.

Service and Installation Pipework

The pipe connecting the consumer's meter with the gas main is known as a service pipe and it should be laid by the shortest practicable route at right angles to the main, with a gradual fall from the meter to the main. The service pipe may be steel or unplasticised polyvynyl chloride. Steel pipes must be protected from corrosion by wrapping with plastic or derso tape, or hessian impregnated with pitch.

Meter Boxes (Fig. 5.6). Whenever possible,

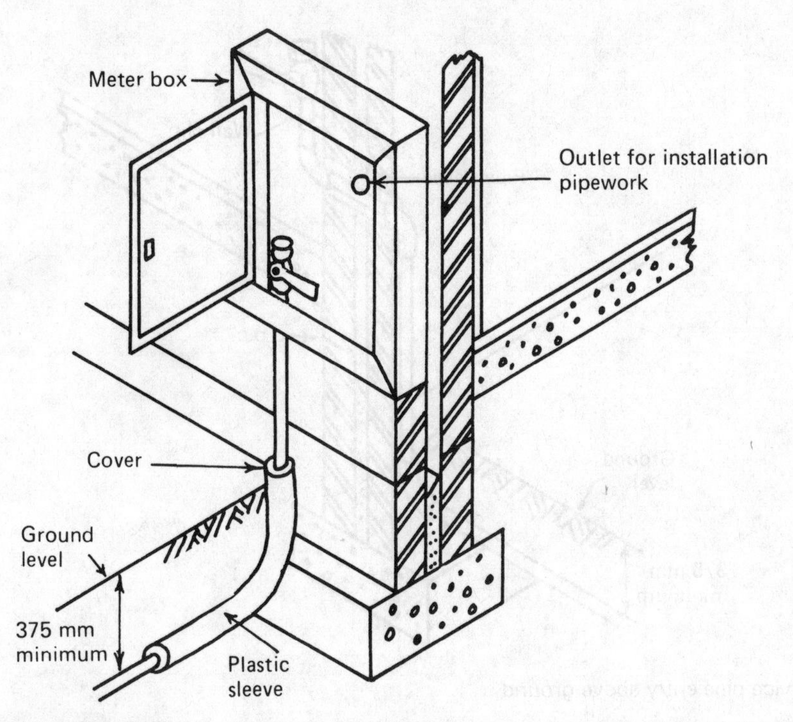

Fig. 5.6 Meter box installation.

the termination of the service pipe should be at an external meter box. Installation time and costs can be reduced by use of this box, which has an external lockable door for meter reading. The box is built into the external leaf of the wall and the service pipe and installation pipework connected to it.

The meter box measures 620 mm by 540 mm and can be painted to match the external paintwork. The external meter box allows the meter to be read without having the inconvenience of making special arrangements for the meter reader or the inconvenience of an estimated account.

Service Pipe Entry Above Ground (Fig. 5.7). If the meter has to be sited inside the building, the gas service pipe may be carried up vertically on the outside wall and pass horizontally through the cavity. A sleeve must be provided where the pipe passes through the wall and after the pipe has been installed both ends of the sleeve should be sealed with cement mortar.

Service Pipe Entry Below Ground. Where construction difficulties preclude the use of an external meter box or riser, the pipe may be carried horizontally through a solid or suspended floor before rising vertically to the meter.

If the pipe passes through a solid floor, a 100 mm diameter sleeve, not more than 2 m long, extending into a 300 mm square hole in the meter position, should be provided (see Fig. 5.8). The annules between the duct and the service pipe should be sealed with a mastic compound. The square hole in the floor should then be filled with sand and the floor finish carried above it. The space around the sleeve on the external wall should be made good with cement mortar.

A service pipe should be laid under a suspended floor only where the void under the floor is ventilated to the outside and the meter positions is not more than 2 m from the outside wall (see Fig. 5.9).

Service Pipe Entry in Garage. The services may be installed in a garage using either the external meter box or a service entry to an internal meter. Where the meter is fitted in the garage, the length of the services inside the garage should not exceed 2 m.

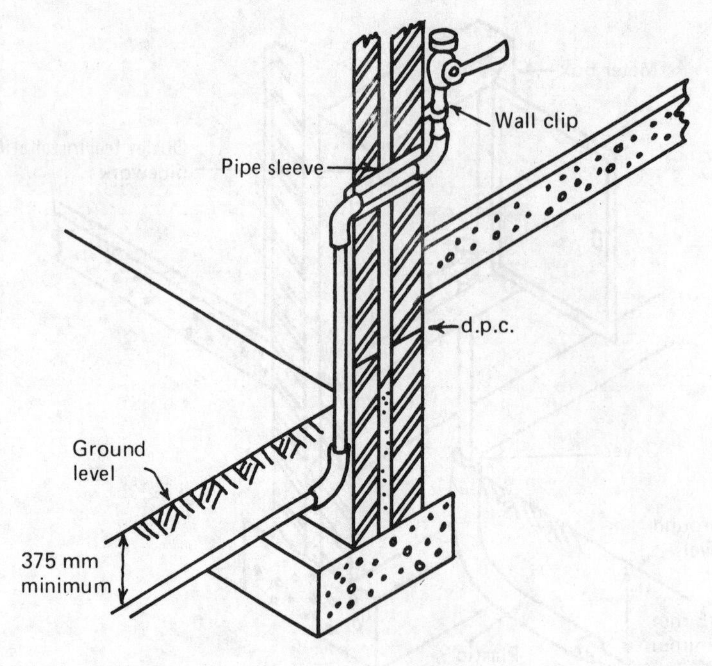

Fig. 5.7 Service pipe entry above ground

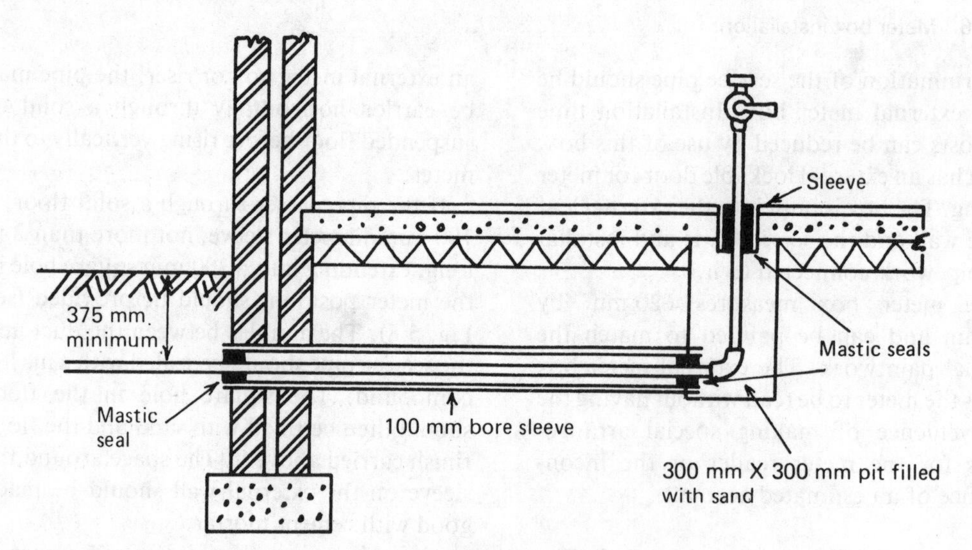

Fig. 5.8 Service pipe below solid floor.

Internal Meters. These should be housed inside a ventilated cupboard, preferably recessed into the wall, and standing about 1.4 m above the floor level. The platform on which the meter stands should be level and the cupboard large enough to permit an air space between the meter and the wall. A main control cock should be situated in the cupboard and, if fitted with a removable key, this should be fixed in a position with a split pin or grub screw.

Meters should not be fitted:

1. under sinks, draining boards, or in a position where they will become wet,
2. in contact with walls,
3. in corrosive or damp atmospheres,

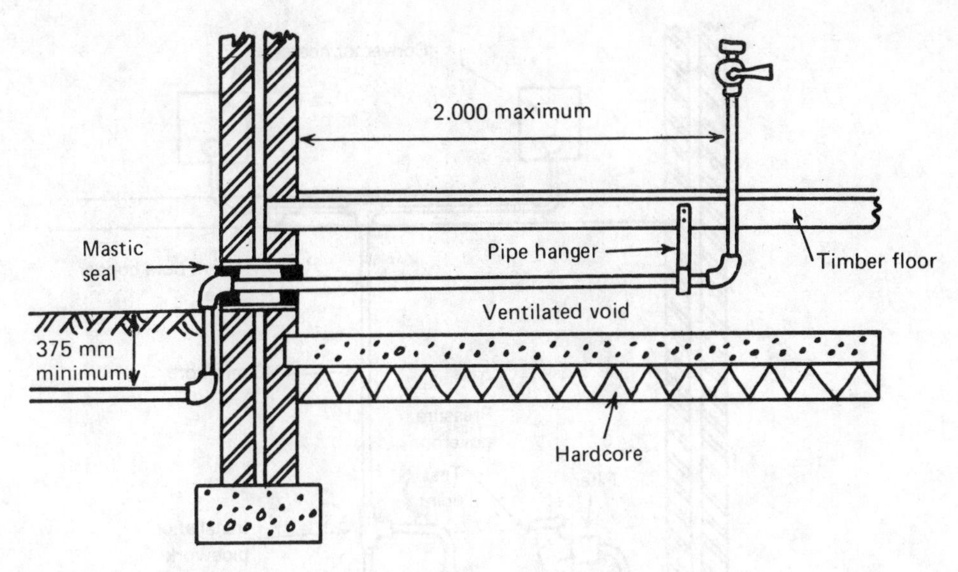

Fig. 5.9 Service pipe below suspended floor.

4. in inaccessible positions,
5. above or near a heating or cooling appliance.

Note: in offices or flats, meters should preferably be located so that the dials can be read through a spyhole in the meter cupboard without the inconvenience of entering the office or flat.

Installation Pipework (Fig. 5.10). The pipework from the meter to the gas appliances is called the installation pipework. Larger diameter pipes are of mild steel, whilst smaller pipes are of copper or brass. Flexible metal pipes are used increasingly for connections to meters and appliances (see Fig. 5.11 and 5.12).

Methods of Installation. It is important to know the exact routing of the pipes to avoid the unnecessary cutting away of joists or the chasing of walls and floors. In general, the pipes should rise steadily from the lowest point and a suitable method is to carry the main installation pipe from the meter and to radiate outwards and upwards from this point. A receiver siphon, tap or plug, placed at the lowest point, will permit the collection and removal of any dust or condensate forming in the pipework.

Points to be observed are as follows.

1. Sharp bends should be avoided.
2. Pipes must never be run inside a cavity or unventilated void space.
3. Pipes should be easily accessible without damage to the structure.
4. Pipes should be of incombustible material and adequately supported.
5. As for service pipes, where an installation pipe passes through a floor or wall, a sleeve should be provided and the space between the sleeve and the pipe filled in.
6. Pipes should not be near to any source of heat or dampness.
7. Each run of pipe should be provided with a means of disconnection for easy replacement.
8. Joints should be of an approved type and jointing compounds used should be noncorrosive and allow easy disconnection.
9. Pipes should not touch other service pipes and should be kept as far away as possible from electrical cables or equipment.
10. Pipes must not be run in bedrooms or bedroom cupboards.

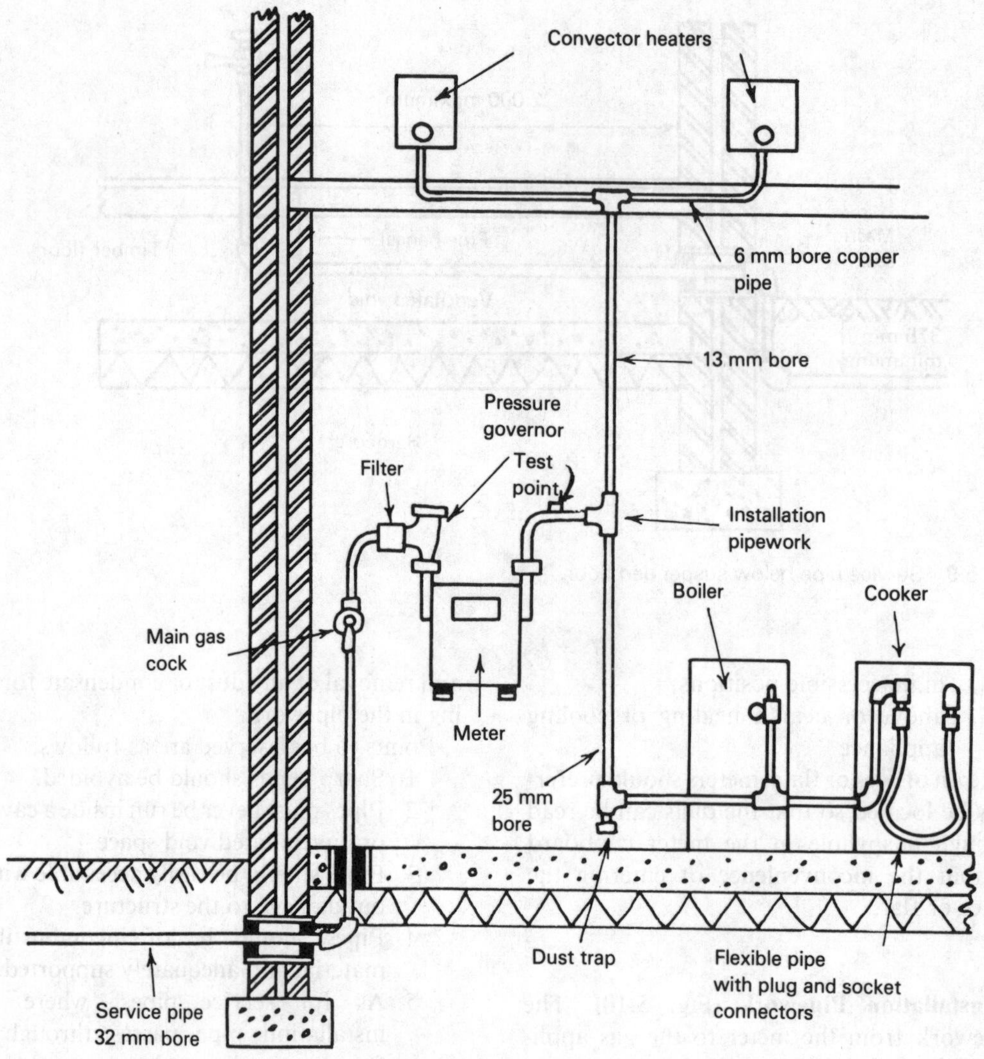

Fig. 5.10 Installation pipework.

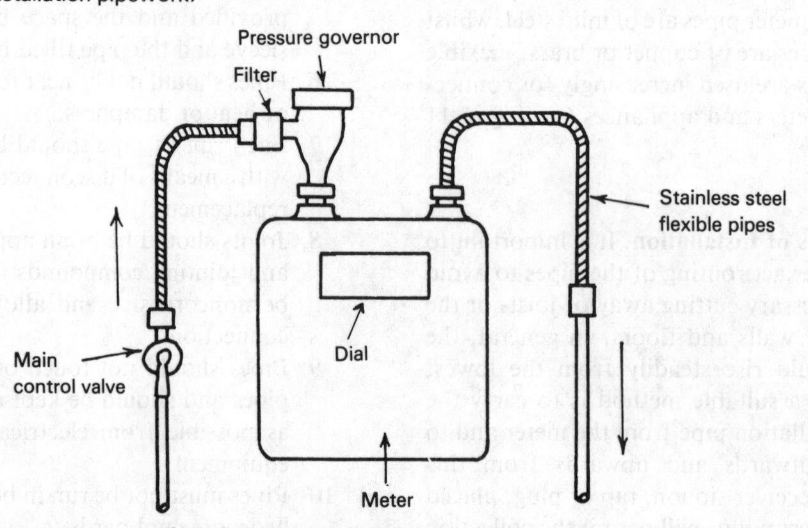

Fig. 5.11 Gas meter connections for domestic buildings.

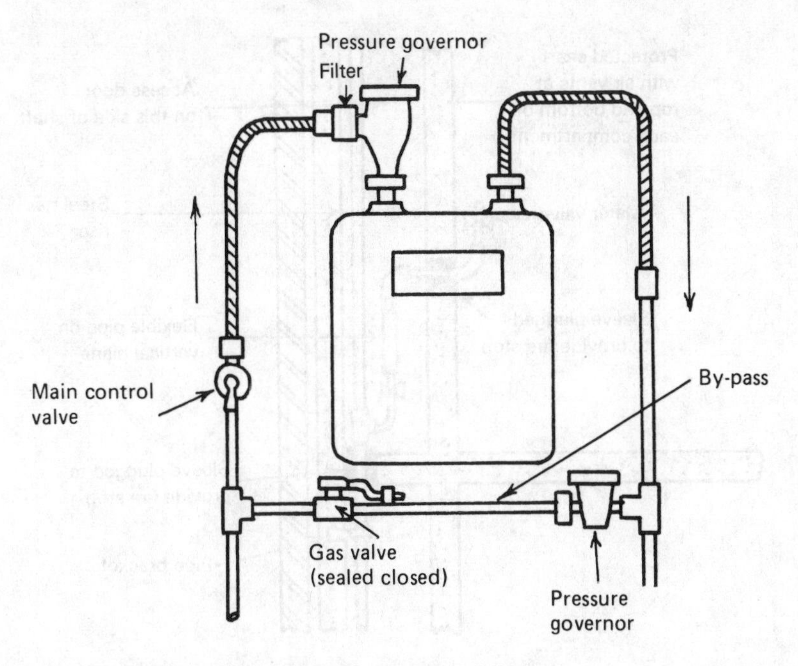

Fig. 5.12 Gas meter connections with by-pass for larger buildings.

High Rise Buildings (Figs. 5.13 and 5.14). The vertical service and installation pipes must only be installed in a protected shaft constructed in accordance with the Building Regulations 1976 Sections E10 and E14.

Alternative methods of constructing the protective shaft include the following.

1. A continuous shaft ventilated to the outside of the building at top and bottom. In this case, a sleeve is required at each point where a horizontal pipe passes through the wall of the shaft and this sleeve must be fire stopped.

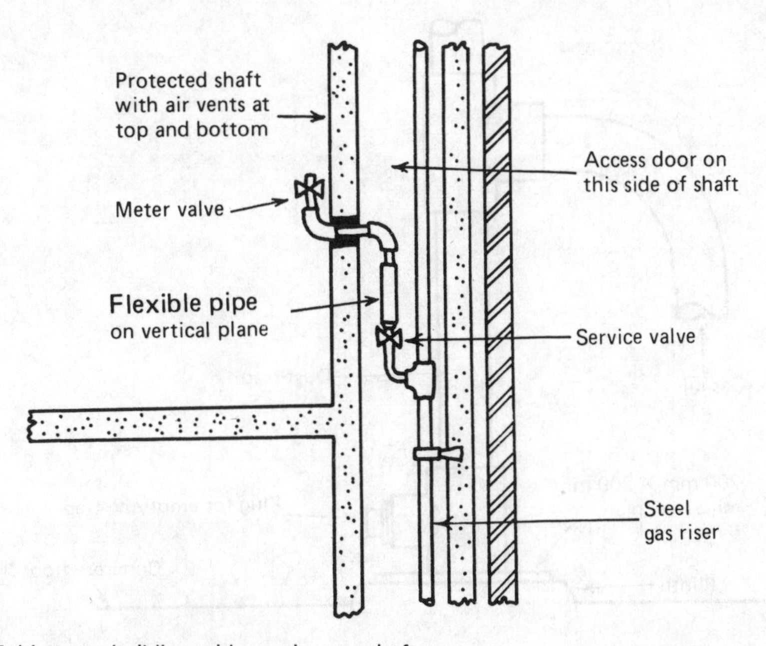

Fig. 5.13 Multi-storey building with continuous shaft.

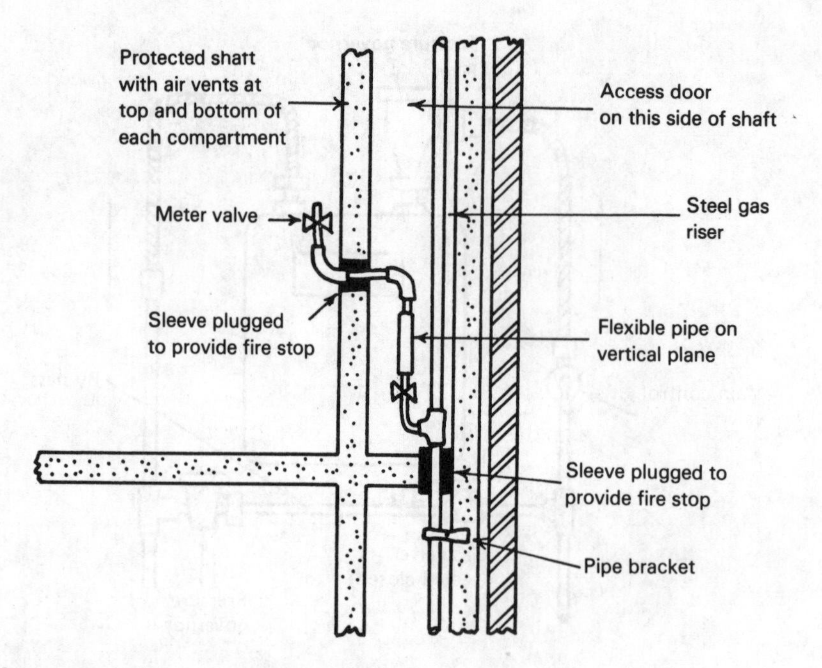

Fig. 5.14 Multi-storey building with sectional shaft.

2. A shaft that is fire stopped at each floor level. In this case, sleeves are required where the riser passes through the floor at each level and where the horizontal pipe passes through the wall of the shaft. The sleeves must be fire stopped. Ventilation to the outside of the building is required at both low and high level for each section of the shaft. Each branch to the meters should be provided with an expansion joint on the vertical plane and the base of the riser should be supported as shown in Fig. 5.15.

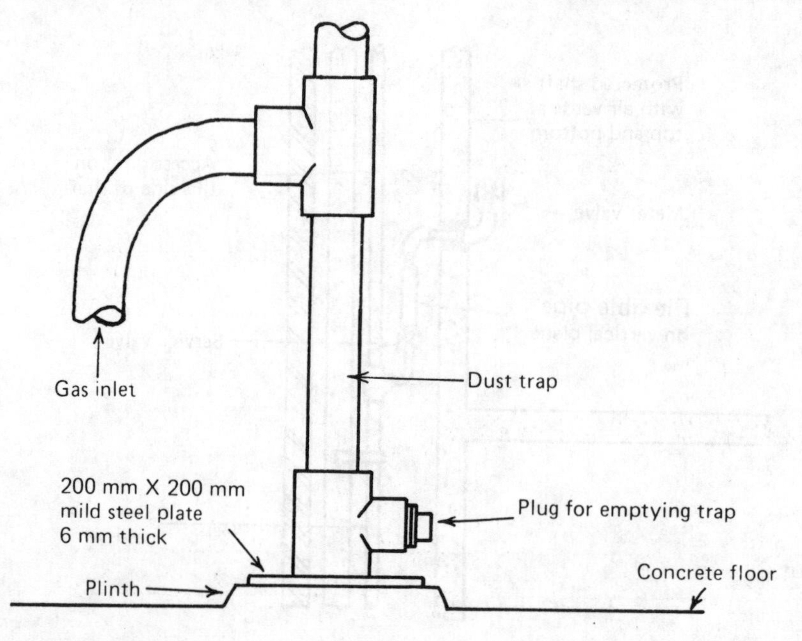

Fig. 5.15 Support for steel riser.

Testing For Soundness (Fig. 5.16)

Whenever new gas installation pipes are installed, either complete systems or additions to existing systems, it is essential that they are tested for soundness before they are covered over or protective coatings applied. The tests are carried out as follows.

(a) New Installations

1. Check that all open ends are closed except one, which is fitted with a T piece having a cock on one branch with a U gauge and pump on the other.
2. Pump air into the pipework until the U gauge shows 300 mm water gauge.
3. Shut the cock and allow 1 min for the air temperature in the pipework to stabilise.
4. Allow a further 2 min and if no pressure drop is registered on the U gauge the system is considered sound.
5. Connect the new installation to the gas supply and pump the air from the pipework by opening the main valve, and test the inlet pipework to the meter by using soapy water on the joints and fittings.

(b) Before connecting an extension to existing pipework, the meter and the existing pipework should be tested as follows.

1. Ensure that the meter and consumer's control have been located to ensure that the whole installation is under test.
2. Turn off all appliances and pilot lights.
3. Turn off the main control valve.
4. Connect the U gauge to the pressure test point on the outlet side of the meter.
5. Open the main control valve until the U gauge registers 200 mm. Adjustment may be made by either releasing gas or pumping air into the pipework.
6. Allow 1 min for the temperature of the air in the pipework to stabilise; leave for a further 2 min and check that the pressure drop does not exceed that shown in Table 1.3, and that there is no smell of gas.

Note: if there is a smell of gas or the pressure drop exceeds that shown in

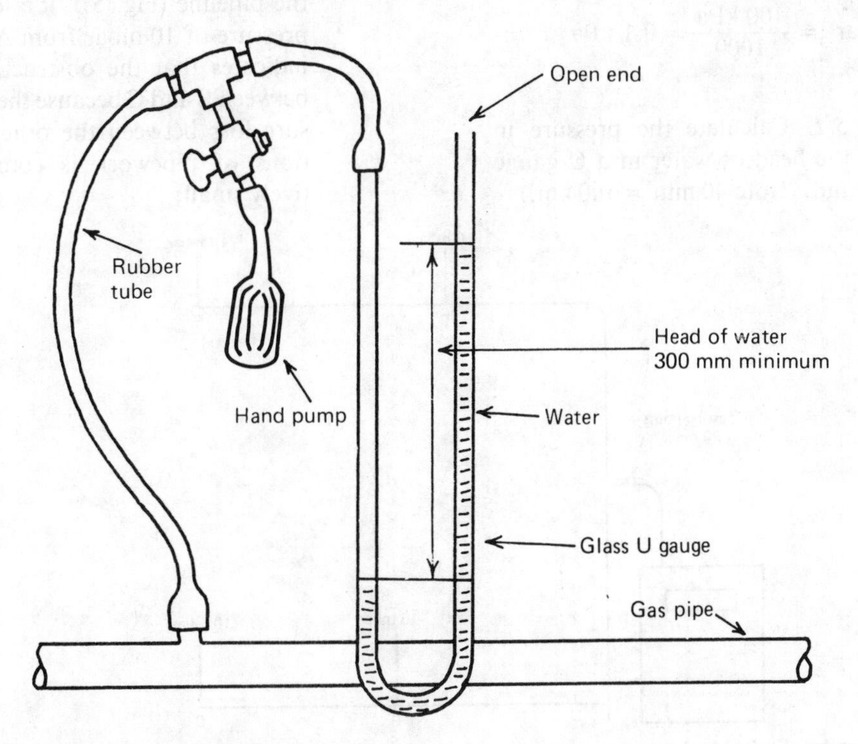

Fig. 5.16 Testing for soundness.

the table, the leak should be located using a sense of smell or soapy water test.

7. Rectify any leaks and repeat the test and if necessary purge the system of air.

Table 5.3 is based on an average length of pipework, but where the existing installation pipework is usually long the permissible pressure drop will be lower. On completion of the test, the meter control valve should be opened and all pilot lights relit.

Calculation of Gas Pressure

When measuring gas pressure with a U gauge (manometer), atmospheric pressure is ignored. The SI unit for pressure is known as the kilopascal (kPa) which is equal to a kN/m^2. In order to find the pressure in kPa, the head of water measured in metres is multiplied by 9.81. The bar, which is not an SI unit, is equal to 100 kPa and the millibar (mbar) is one thousandth part of a bar. Therefore

$$1 \text{ mbar} = \frac{100 \text{ kPa}}{1000} = 0.1 \text{ kPa}$$

Example 5.1. Calculate the pressure in mbar when the head of water in a U gauge registers 40 mm. (Note 40 mm = 0.04 m.)

Table 5.3 Maximum Permissible Pressure Drop during Two Minutes Final Test Period for Existing Installations

Type of meter	Pressure drop	
	mm water gauge	mbar (approx.)
U6 (D07)	40	4.0
P1 (D1)	25	2.5
P2 (D2)	15	1.5
P4 (D4)	5	0.5

$$\text{Pressure} = 0.04 \times 9.81 \times 10 = 3.924 \text{ mbar}$$

Manometers may be obtained which give a direct reading in millibars instead of millimetres. Other uses of the manometer are as follows.

1. Locating excessive pressure loss due to:

   (a) partial stoppage, by taking working pressures reading along the pipeline (Fig. 5.17): a loss of pressure of 10 mbar from A to C indicates that the obstruction is between B and C because the pressure loss between the other sections of pipework is comparatively small;

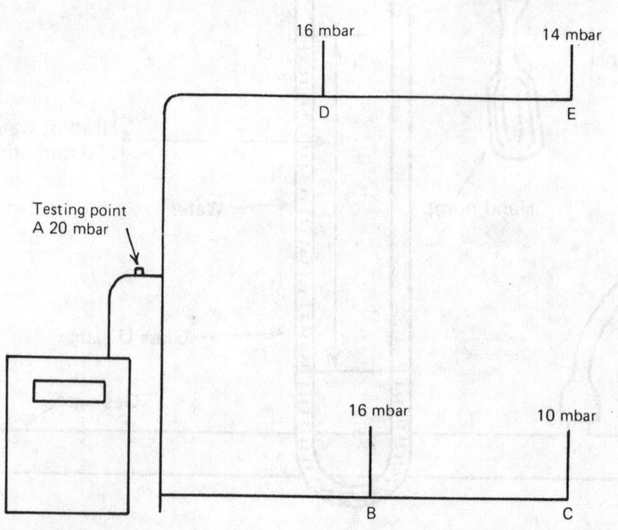

Fig. 5.17 Test for partial blockage.

(b) inadequate pipe size by comparing the difference in readings between the inlet and outlet pressures.

2. Detecting overloading of a meter. This may be found by taking the difference in readings between the inlet and outlet pressure, which difference should not exceed 1.25 mbar.
3. Adjusting appliances. Since gas appliances are designed to give their best performance at a certain pressure and gas flow rate, those can be checked by means of the manometer in conjuction with the pressure governor on the appliance. After setting the pressure, the gas rate to the appliance can be checked by timing the test dial on the meter with a watch.
4. Checking continuity. The gauge can be attached to various points on the pipeline to see if there is any undue drop in pressure. Note: the Gas Safety Regulations 1972 require meter pressure governors to be sealed.

Exercises

1. Describe the formation of natural gas.
2. State the main characteristics of natural gas.
3. Sketch and describe the type of burner required for natural gas.
4. Define the following terms:
   (a) burning velocity,
   (b) aerated burner,
   (c) nonaerated burner.
5. Sketch the method of connecting the gas service pipe to the main and state the types of material used for service pipes.
6. Sketch the method of connecting a service pipe to the meter when the meter is sited
   (a) in an external meter box,
   (b) in an internal cupboard.
7. State the safety consideration that must be made when deciding upon the position of a gas meter inside a building.
8. Sketch the method of installing meters and gas riser pipework in high-rise buildings.
9. State the points that must be observed when installing gas pipework in buildings.
10. Sketch and describe the method of carrying out a test for soundness for
    (a) new gas installations,
    (b) existing gas installations.
11. If a U gauge connected to a gas installation shows a reading of 30 mm water gauge, calculate the pressure of gas in the pipework in millibars and pascals.

Chapter 6
Gas Controls

Gas Cocks

The simplest and most common type of control is the gas cock, which is used to shut off the supply of gas to the meter or to individual gas appliances. The cock consists of a tapered plug having a slot through it, which fits into a tapered body. The two connecting surfaces of the plug and tapered body are machined to give a gas-tight seal and these surfaces should be smeared with grease to act as a lubricant. The cock is opened or closed by turning the plug through an angle of 90°. Fig. 6.1 shows a section through a main control cock which incorporates a union to facilitate the removal of the meter.

For safety reasons, a drop-fan cock may be used as shown in Fig. 6.2. The fan is hinged to the plug and has lugs which hold the fan upright when the gas is on. When the cock is turned off, the fan falls and the lugs engage with a slot in the tapered body of the cock. The plug in this off position cannot be turned on until the fan is deliberately held upright and this prevents the cock from being turned on accidentally.

An even safer type of control is known as the plug-in safety cock, shown in Fig. 6.3. This type of cock cannot be turned on without first inserting the outlet pipe to the body of the cock. When the outlet pipe plug is inserted and turned to engage the lugs, two pins on the plug also engage a groove, and a gas-tight seal is made between the two components by compressing the main spring, which forces the two conical surfaces together.

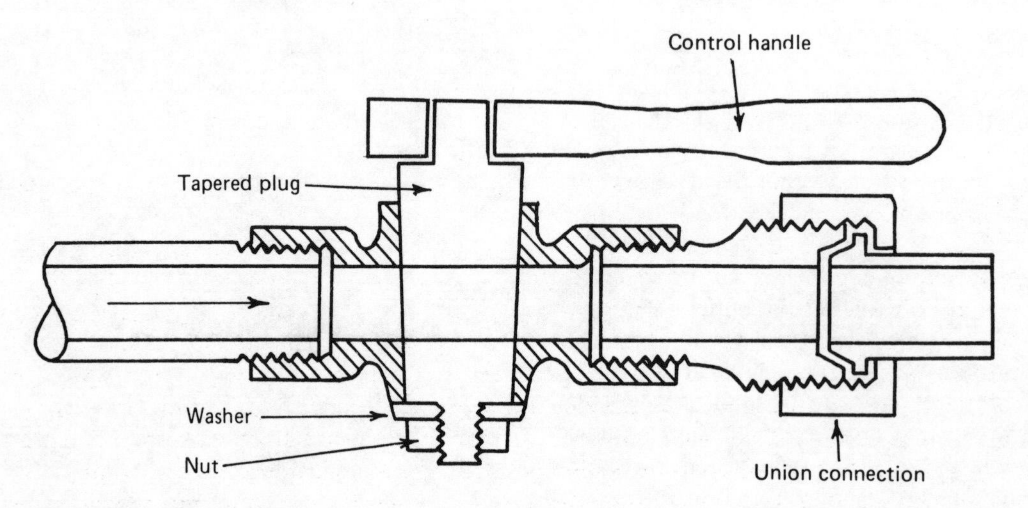

Fig. 6.1 Main control cock.

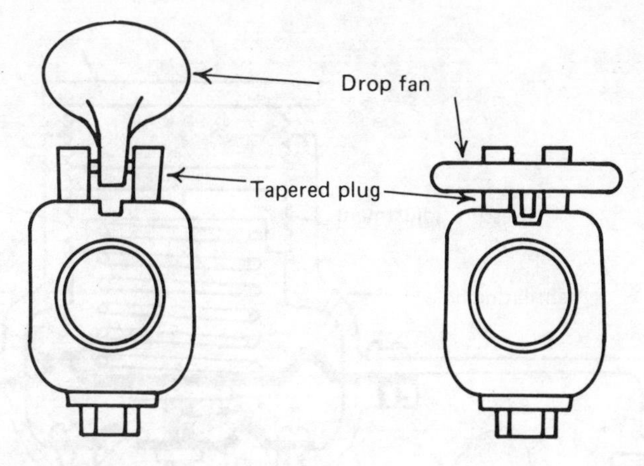

Fig. 6.2 Drop fan safety cock.

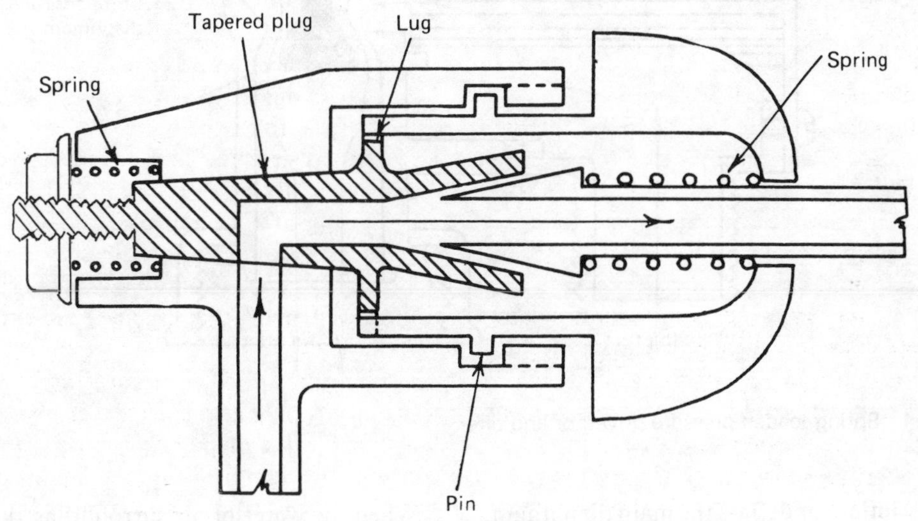

Fig. 6.3 Plug-in safety cock.

Pressure Governor

The control of the gas pressure at the meter or appliances may be achieved by use of a constant pressure governor fitted on the inlet pipe. The governor may be loaded by lead weights or a spring, and the extent of the loading may be adjusted to provide the correct gas pressure at the outlet. A weight-loaded governor must always be in a horizontal position so that the weights act vertically on the diaphragm. Spring-loaded governors, however, may be fixed in any position provided it is easy to adjust the tension on the spring. Fig. 6.4 shows a section through a spring-loaded governor which operates as follows.

1. Gas enters the governor at the inlet pressure and passes through valve A and also through the space B between the two diaphragms.
2. The main diaphragm is loaded by the spring and the upward and downward forces of gas acting on the diaphragm are balanced. These equal forces have a stablising effect on valve A and prevent any tendency to oscillation.
3. Any fluctuation of inlet pressure will

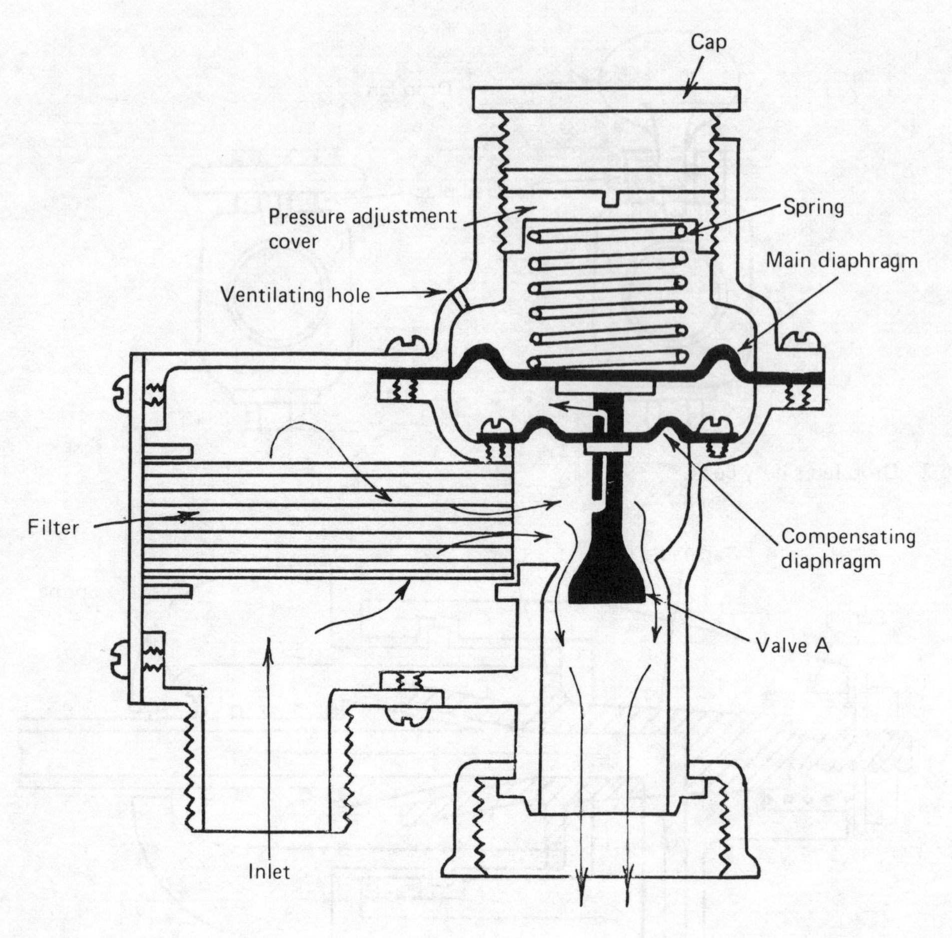

Fig. 6.4 Spring-loaded pressure governor and filter.

inflate or deflate the main diaphragm, thus raising or lowering valve A, altering the resistance to the flow of gas and ensuring a constant pressure at the outlet.

Thermostats

A thermostat is a device that opens or closes a gas valve by sensing the temperature surrounding it. The sensing of the temperature may be achieved by use of the differential expansion of metals, or of the expansion of liquids or vapours when heated.

Rod-Type Thermostat (Fig. 6.5). This type of thermostat may be used for gas-fired boilers, cooker ovens and water heaters. When the water or air surrounding the brass tube becomes hot, it expands and carries the invar rod (which expands very little) with it. This brings the gas valve closer to its seating, thus reducing or shutting off the flow of gas to the burner. When the water or air cools down, the brass tube contracts, carrying with it the invar rod, thus opening the valve. The bypass permits a small amount of gas to flow when the valve is closed and maintains a small flame at the pilot.

Vapour Expansion Thermostat (Fig. 6.6). These are often used for water heaters and space heaters. The thermostat consists of a capillary tube, probe and bellows, filled with a heat-sensitive vapour such as ether. The thermostatic probe senses the temperature of the water or air surrounding it and the vapour

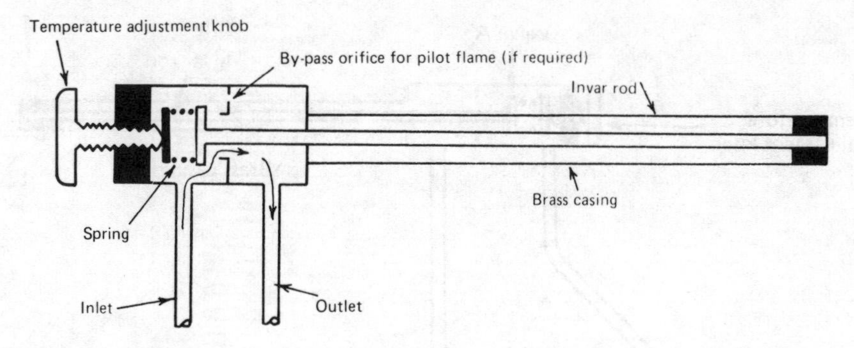

Fig. 6.5 Rod-type thermostat.

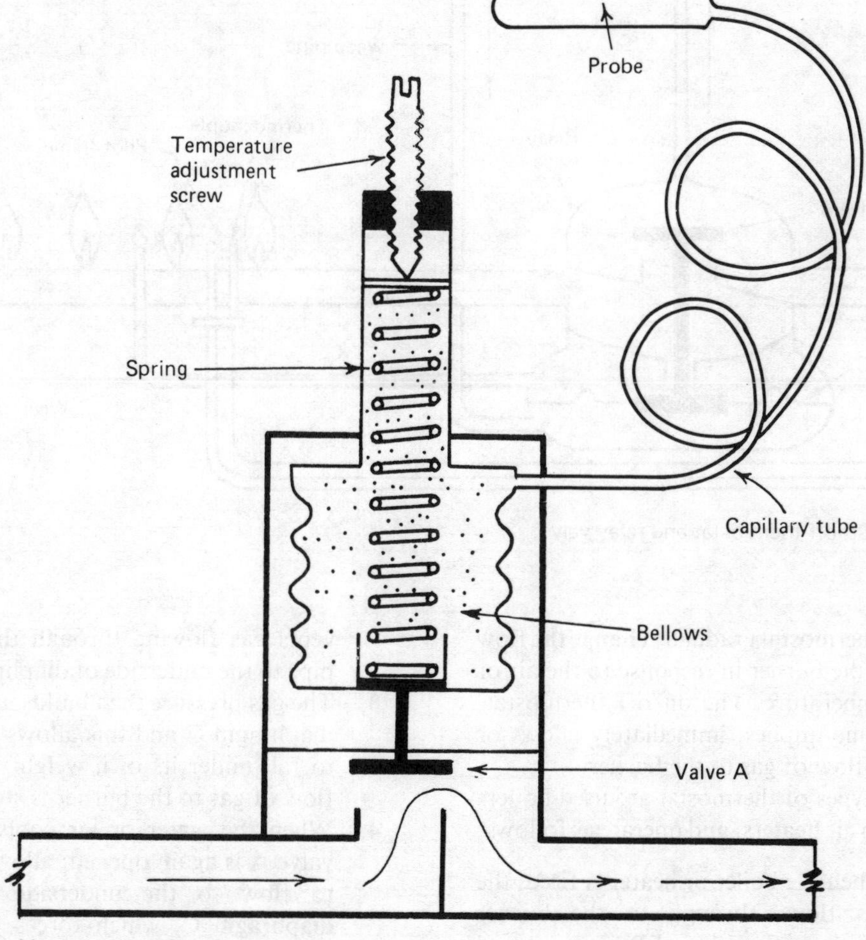

Fig. 6.6 Vapour expansion thermostat.

inside expands. This expansion of the vapour causes the bellows to expand which, in turn, forces the valve A closer to its seating, thus gradually preventing the flow of gas to the burner. When the water or air temperature surrounding the probe lowers, the vapour contracts, which in turn allows the bellows to contract and valve A to open, thus allowing gas to flow again to the burner.

On/Off Thermostat and Relay Valve (Fig. 6.7). The rod and vapour expansion

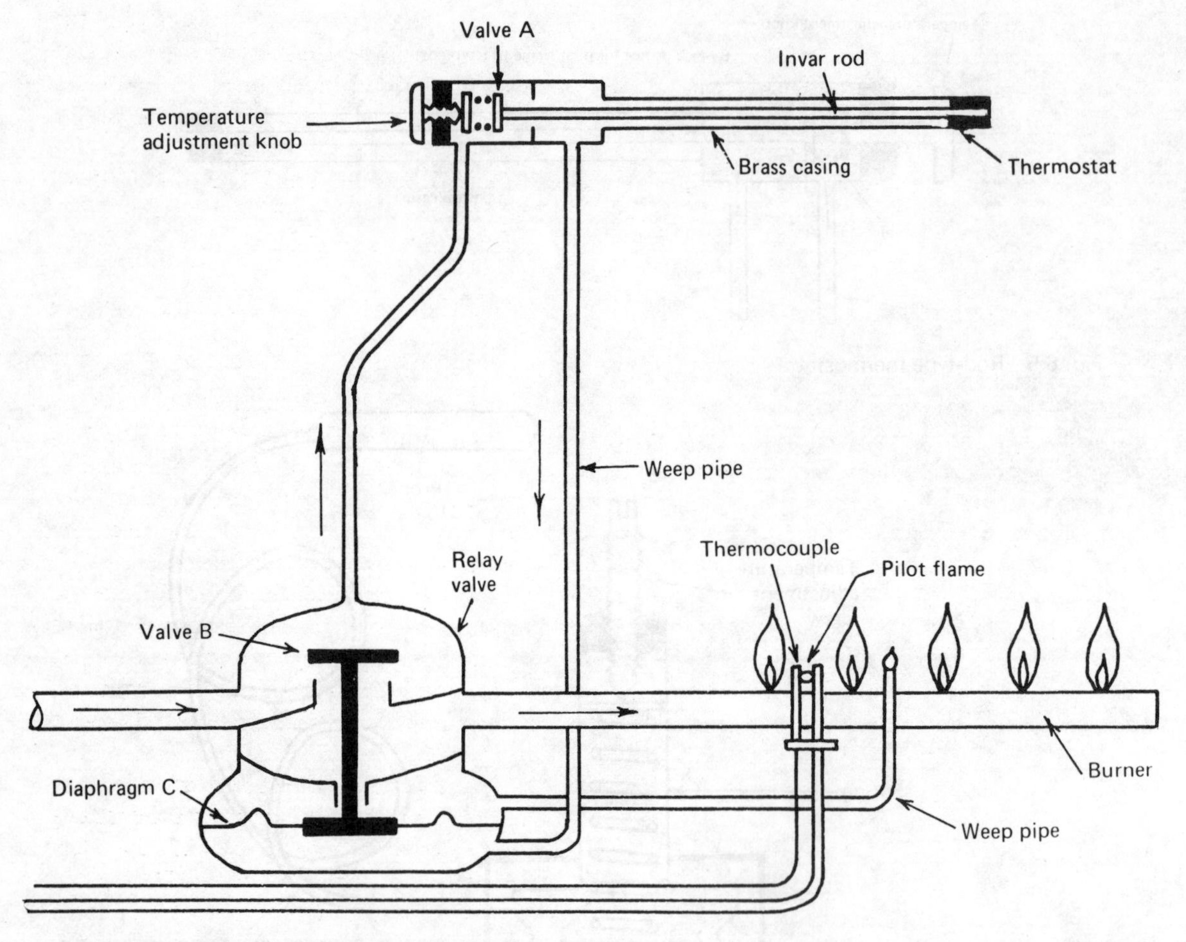

Fig. 6.7 On-off thermostat and relay valve.

types of thermostat gradually change the flow of gas to the burner in response to the air or water temperature. The on/off thermostat, as the name implies, immediately allows or stops the flow of gas to the burner.

These types of thermostat are used boilers and warm air heaters, and operate as follows.

1. When the boiler or heater is fired, the gas flows through to the burner because valves A and B are open and the gas pressures above and below diaphragm C are equal.
2. When the water or air reaches the required temperature, the brass casing expands sufficiently to draw valve A, connected to the invar rod, to a closed position. The closing of valve A prevents gas flowing through the weep pipe to the underside of diaphragm C.
3. The gas pressure then builds up above diaphragm C and this allows valve B to fall under its own weight and the flow of gas to the burner is stopped.
4. When the water or air cools down, valve A is again opened, allowing gas to flow to the underside of the diaphragm C, which forces valve B open, and gas is again allowed to flow to the burner.

Note: for valve B to be forced open, the gas above diaphragm C must be removed. This is achieved by a bypass orifice and a weep pipe to the burner, through which the gas escapes and is ignited and burned.

Flame-Failure Devices

It is essential to prevent unburnt gas reaching the burner of an automatic gas appliance in the event of the pilot flame being extinguished. To prevent this possibility and risk of an explosion, a flame-failure device must be fitted.

Mercury Vapour Flame-Failure Device (Fig. 6.8). Mercury is contained in the sensing element, A, which is connected to diaphragm B by a capillary tube. When the sensing element is exposed to the heat of the pilot flame, the mercury vaporises. The resultant pressure of the vapour causes the diaphragm B to expand and this movement is relayed through lever C to valve D, which is forced open and allows gas to flow to the burner.

Should the pilot flame be extinguished, the sensing element A cools and the mercury vapour liquefies, allowing diaphragm B to contract. Spring E then forces valve D down on to its seating, thus preventing the flow of gas to the burner.

Bi-metal Strip Flame-Failure Device (Fig. 6.9). This device consists of two alloys fixed rigidly together and bent into a U shape. The alloys have widely different coefficient of expansion and when the strip is heated, the brass on the outside expands more than the invar on the inside, causing the strip to deflect inwards. This movement opens the gas valve and allows gas to flow to the burner, where it is ignited by the pilot flame.

Should the pilot flame the extinguished, the bi-metal strip cools and deflects outwards, thus closing the gas valve.

Thermoelectric Flame-Failure Device (Fig. 6.10). The sensing element of this device is a thermocouple, which consists of two dissimilar metals joined together to form an electrical circuit. One junction of the two metals is heated by the pilot flame and is known as the 'hot junction' and the other remains relatively cool and is known as the 'cold junction'. When this temperature differential is established, a small electric current is generated and this energy is used to operate an electromagnet in the valve. The electromagnet holds the safety cut-out valve in an open position and allows gas to flow to the burner.

Should the pilot flame be extinguished, the thermocouple is prevented from producing a small current of electricity and the electromagnet is no longer energised. The cut-out valve spring forces the cut-out valve on to its seating and this prevents gas from flowing to the burner.

To light the burner, the following

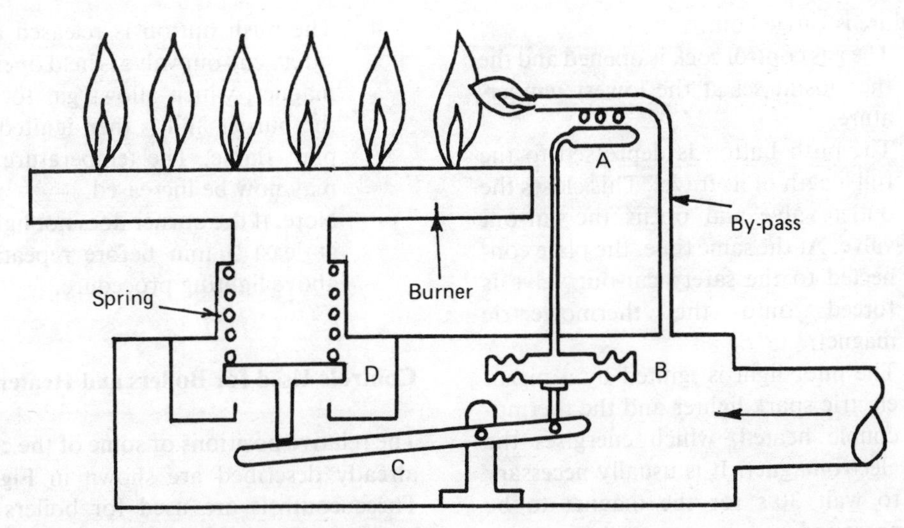

Fig. 6.8 Mercury vapour flame-failure device.

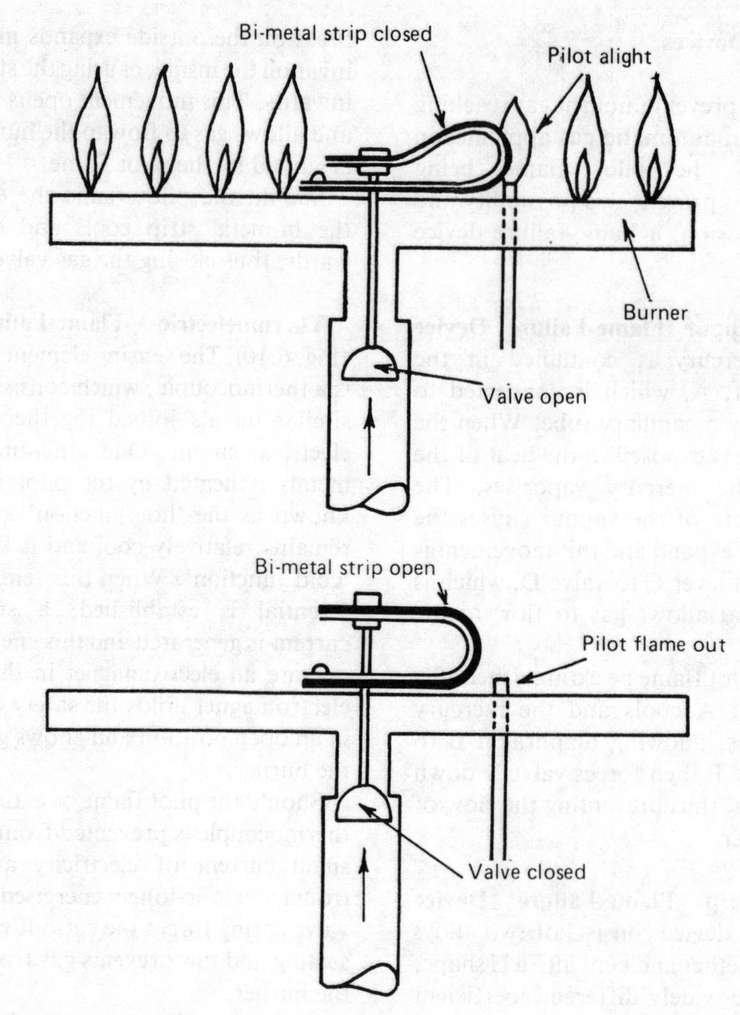

Fig. 6.9 Bi-metal strip flame-failure device.

procedure is carried out.

1. The gas control cock is opened and the thermostat set at the lowest temperature.
2. The push button is depressed to the full length of its travel. This closes the outlet valve and opens the cut-out valve. At the same time, the plate connected to the safety cut-out valve is forced onto the thermoelectric magnet.
3. The pilot light is ignited by a piezoelectric spark lighter and the thermocouple heated, which energises the electromagnet. It is usually necessary to wait 30 s for the magnet to be energised.
4. The push button is released and the safety cut-out valve is held open by the magnet, which allows gas to flow to the burner and is then ignited by the pilot flame. The temperature setting may now be increased.
   Note: if the burner does not light, wait at least 3 min before repeating the above lighting procedure.

Controls Used for Boilers and Heaters

The relative positions of some of the controls already described are shown in Fig. 6.11. These controls are used for boilers or air heaters and are usually sited inside the casing,

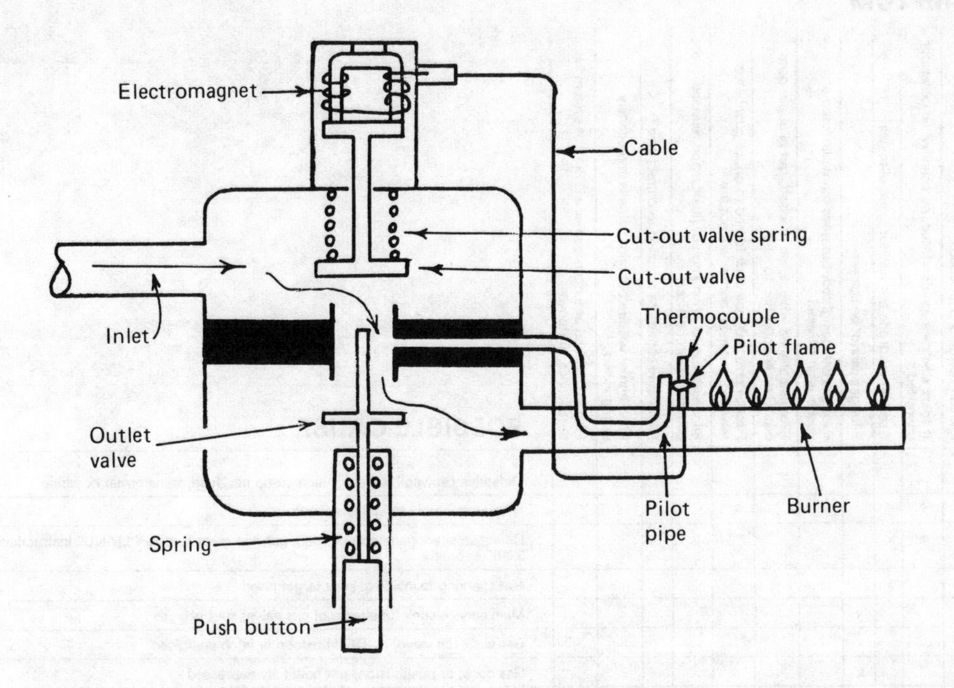

Fig. 6.10 Thermoelectric flame-failure device.

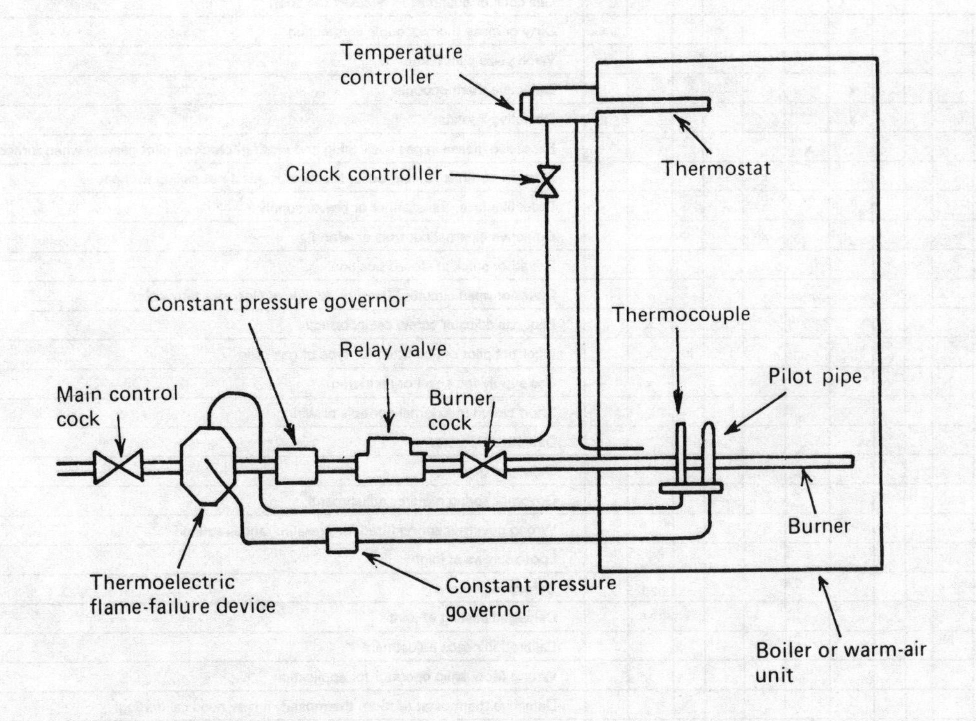

Fig. 6.11 Controls used for heaters.

Table 6.1 Gas Valves — Fault Finding Guide

POSSIBLE CAUSE	If Glowcoil ignition is fitted, Glowcoil will not light when ignition button is depressed	If spark electrode ignition fitted, no spark when generator energised	Spark at electrode but will not ignite pilot gas	Pilot gas cannot be lighted	Pilot gas can be lighted, but goes out when knob or pushbutton released	Pilot established when knob released, but pilot gas goes out when knob turned to 'ON'	Pilot established, gas cock turned to on, or pushbutton released, but main burner will not fire	Pilot established when knob or pushbutton released, but pilot flame wrong size	Pilot flame cannot be adjusted to correct size	Main burner will fire but pilot flame is immediately extinguished	Pilot established and main burner will fire but system liable to nuisance shutdown
Defective Glowcoil, other ignition comp'nts, fuse, transformer or wiring	x										
No power supply to ignition circuit	x	x									
Defective spark generator or spark gap too great (see APPLIANCE instructions or set to 3/16" approx.)		x	x								
Not sparking to inside of pilot target hood			x								
Main service cock (upstream of gas valve) shut off				x							
Gas cock (on valve) in 'OFF' position or latch energised				x							
Gas cock, or pushbutton, not held fully depressed				x							
Pilot gas line not yet purged of air				x							
Pilot filter blocked				x							
Pilot gasways blocked				x							
Pilot gas adjuster closed too far				x							
Pilot gas line blocked				x							
Pilot burner orifice blocked				x							
Gas cock or pushbutton released too soon					x						
Dirty or loose thermocouple connection					x						x
Wrong size pilot flame					x					x	x
Defective thermocouple					x						x
Defective Pilotstat					x						x
Excessive grease in gas cock (plug and seating) blocking pilot gasway when turned						x					
External controls (Clock, boiler thermostat, etc.) not calling for heat							x				
Defective fuse, transformer or power supply							x				
Defective external controls or wiring							x				
Governor stuck in closed position							x				
Operator open circuited (test for voltage at Operator terminals)							x				
Pilot gas adjuster screw set incorrectly								x		x	
Incorrect pilot orifice fitted for type of gas used									x	x	
Gas supply too small or restricted										x	
Short circuit in external controls or wiring											
Defective Operator											
Defective governor											
Governor spring requires adjustment											
Wrong governor spring fitted for pressure range required											
Loose screws at joint											
Defective gasket											
Damaged casting at joint											
Calibration needs adjustment											
Wrong Modusnap operator for application											
Defective thermostat (if room thermostat, it may need calibrating)											
Poor location for room thermostat											
Wrong low fire orifice fitted or incorrect adjustment											
Leak at pressure regulator or Operator gasket											

(Column headings above are listed under SYMPTOM.)

POSSIBLE CAUSE	Main gas will not shut down in response to external controls	Gas valve noisy when main burner is firing	Main burner pressure cannot be maintained at constant level	Main burner pressure incorrect	Gas leaks from joint at gas valve body/governor/Pilotstat/operator on soap solution check	(V5130 only) Correct temperature settings cannot be obtained	(Modusnap valves only) Correct temperature settings cannot be obtained	(Electric Operator valves only) Correct temperature settings cannot be obtained	(Modusnap valves only) Low fire gas rate incorrect	(Modusnap valves only) Valve will not shut down when heat demand falls below low fire rate	(Servo-series valves only) Electrical power at Operator terminals but valves will not open
Defective Glowcoil, other ignition comp'nts, fuse, transformer or wiring											
No power supply to ignition circuit											
Defective spark generator or spark gap too great (see APPLIANCE instructions or set to 3/16" approx.)											
Not sparking to inside of pilot target hood											
Main service cock (upstream of gas valve) shut off											
Gas cock (on valve) in "OFF" position or latch energised											
Gas cock, or pushbutton, not held fully depressed											
Pilot gas line not yet purged of air											
Pilot filter blocked											
Pilot gasways blocked											
Pilot gas adjuster closed too far											
Pilot gas line blocked											
Pilot burner orifice blocked											
Gas cock or pushbutton released too soon											
Dirty or loose thermocouple connection											
Wrong size pilot flame											
Defective thermocouple											
Defective Pilotstat											
Excessive grease in gas cock (plug and seating) blocking pilot gasway when turned											
Extenal controls (clock, boiler thermostat, etc.) not calling for heat)											
Defective fuse, transformer or power supply											
Defective external controls or wiring											
Governor stuck in closed position											
Operator open circuited (test for voltage at Operator terminals)											
Pilot gas adjuster screw set incorrectly											
Incorrect pilot orifice fitted for type of gas used											
Gas supply too small or restricted											
Short circuit in external controls or wiring	x										
Defective Operator	x	x								x	
Defective governor			x								
Governor spring requires adjustment				x							
Wrong governor spring fitted for pressure range required				x							
Loose screws at joint					x						
Defective gasket					x						
Damaged casting at joint					x						
Calibration needs adjustment						x	x				
Wrong Modusnap operator for application							x				
Defective thermostat (if room thermostat, it may need calibrating)								x			
Poor location for room thermostat								x			
Wrong low fire orifice fitted, or incorrect adjustment									x		
Leak at pressure regulator or Operator gasket											x

(Column headings: **SYMPTOM**)

although for clarity they are shown outside in the figure. The constant pressure governor, thermoelectric flame-failure device and relay valve are often combined into one unit.

Thermal Cut-Off Device (Fig. 6.12)

These are used to shut off the supply of gas automatically in the event of fire. When the temperature surrounding the device reaches a temperature of 110 °C, the metal holding a spring-loaded cylinder melts, and the cylinder is forced by the spring onto a seating, closing the valve.

Table 6.1 gives a gas valve fault-finding guide which will be useful when carrying out maintenance work.

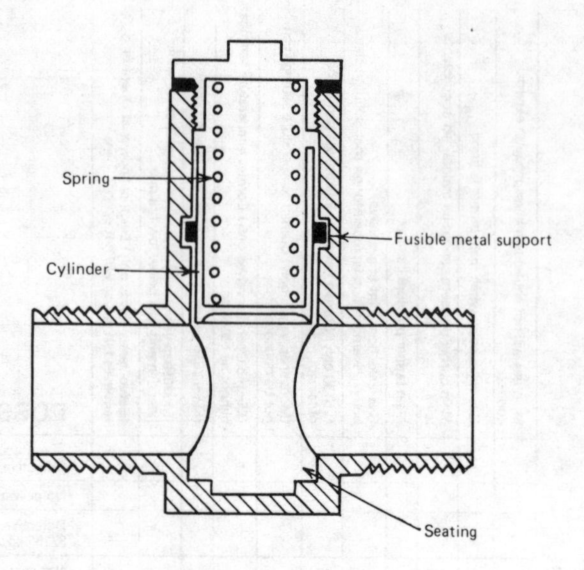

Fig. 6.12 Thermal cut-off device.

Exercises

1. Sketch sections and describe the operation of
   (a) a gas cock,
   (b) a drop-fan cock,
   (c) a pressure governor.
2. Sketch and describe the operation of
   (a) a rod-type thermostat,
   (b) a vapour-expansion-type thermostat.
3. Describe the operation of an on/off thermostat and relay valve.
4. Explain the operation of the following types of device
   (a) a mercury vapour flame-failure device,
   (b) a bi-metal strip flame-failure device.
5. Sketch and describe the operation of a thermoelectric flame-failure safety device. Explain the procedure that must be carried out with this type of device in order to light the burner.
6. Sketch an elevation showing the relative positions of the various controls required for gas boilers or gas warm air heaters.
7. Sketch a section through a thermal cut-off device and state its operating temperature.

Chapter 7
Gas Flues and Fires

Definitions

The Building Regulations 1976 uses the following.

APPLIANCE: a heat producing appliance (including a cooker) designed to burn gaseous fuels.

APPLIANCE VENTILATION DUCT: a duct forming a passage which in one part serves to convey combustion air to one or more gas appliances, in another part serves to convey the products of combustion from one or more gas appliances to the external air and intermediately serves both purposes.

BRANCH (SHUNT) FLUE SYSTEM: a flue system comprising a common main flue into which more than one conventionally flued appliance may discharge on each or several floors in a medium-rise or multi-storey building.

CHIMNEY: includes any part of the structure of a building forming part of a flue other than a flue pipe.

CLASS 11 GAS APPLIANCE: a gas appliance having an input rating not exceeding 45 kW, or an incinerator having a refuse chamber not exceeding 0.03 m^3 in volume.

DISCHARGE: the discharge of the products of combustion.

FLUE: a passage for conveying the discharge of an appliance to the external air, including any part of the passage in an appliance ventilation duct which serves the purpose of a flue.

GAS FIRE: a flued appliance for heating one room, mainly by the emission of radiant heat, and not comprising any water-heating component.

HIGH-RATING APPLIANCE: a gas appliance having an input rating exceeding 45 kW, or an incinerator having a refuse combustion chamber exceeding 0.08 m^3 in volume.

INSULATED METAL CHIMNEY: a chimney comprising a flue lining, noncombustible thermal insulation, and a metal outer casing.

MAIN FLUE: a flue serving one appliance.

PERMANENT VENT: a purpose-made opening or duct designed to allow the passage of air at all times.

ROOM-SEALED APPLIANCE: a gas appliance that draws in combustion air from a point immediately adjacent to a point where it discharges its products of combustion and is so designed that the inlet, outlet and combustion chamber of the appliance, when installed, are isolated from the room or internal space in which the appliance is situated, except for a door for ignition purposes.

SUBSIDIARY FLUE: a flue conveying the discharge of an appliance into a main flue.

TERMINAL: a device, fitted at the termination of the chimney or flue, designed to allow free egress of the products of combustion, to minimise down draught and to prevent the entrance of foreign matter which may cause restriction of the flue.

VENTILATION OPENING: any openable part of a window or hinged panel, adjustable louvre or other means of ventilation that opens directly into the external air, but excluding any opening associated with mechanically operated systems.

Sizes of Gas Flues (open-flued appliances)

The Building Regulations 1985 give the following sizes of flues:

1. No dimension should be less than 63 mm, or
2. flues for decorative appliances, i.e. log and other solid fuel fire effect gas appliances, should have no dimension across the axis less than 175 mm, or
3. flues for gas fires should have a cross-sectional area of at least 12 000 mm^2 and if the flue is rectangular, the greater dimension should not be more than six times the lesser, or
4. any other appliance should have a flue with a cross-sectional area at least equal to the size of the outlet from the appliance and if the flue is rectangular, the greater dimension should not be more than five times the lesser.

Materials used for Flue Pipes

The Building Regulations 1985 specify the following

1. sheet metal as described in BS715 : 1970, sheet metal flue pipes and accessories for gas fired appliances, or
2. stainless steel, or
3. asbestos cement as described in BS567 : 1973 (1984, 'Asbestos-cement flue pipes and fittings, light quality', or
4. cast iron as described in BS41 : 1973 (1981) 'Cast iron spigot and socket flue or smoke pipes and fittings', or
5. flue blocks as described in BS1289 : 1975, 'Precast concrete flue blocks for domestic gas appliances'.

A chimney may be built of any masonry material with a lining or of flue blocks without a lining. Linings may be of one of the following:

1. clay flue liner with rebated or socketed joints as described in BS1181 : 1971 (1977), 'Clay flue linings and flue terminals'
2. imperforate clay flue pipes as described in BS65 : 1981, 'Specification for vitrified clay pipes fittings and joints'.
3. a flexible flue liner may be used in a chimney if:
   (a) the liner complies with the requirements of BS715 : 1970, 'Sheet metal flue pipes and accessories for gas fired appliances', and
   (b) the chimney was built before 1 February 1966.

Note: Flue pipes with spigot and socket joints should be fitted with sockets uppermost and flue linings should be fitted with the sockets or rebates uppermost to prevent condensate running out and to prevent any caulking material from being adversely affected. Joints between the liners and masonry should be filled with mortar. Factory-made insulated chimneys may also be used.

Outlets from Flues

The outlet should be:

1. so situated externally that air may pass freely across it at all times, and
2. at least 600 mm from any opening into the building, and
3. fitted with a flue terminal where any dimension across the axis of the flue outlet is less than 200 mm.

Openings into Flues

No opening should be made in a flue except for:

1. inspection or cleaning, when the opening should be fitted with a non-combustible gas-tight cover, or
2. a draught diverter.

Communication Between Flues

No flue should have openings to more than one room or space except for an opening for inspection or cleaning.

Purposes of Flues

Gas flues in a building provide two main functions, to

1. remove the products of combustion from the appliance to the atmosphere,
2. assist in providing ventilation of the room where the appliance is installed.

Design Factors

The decisive requirements for a flue are:

1. gas rate per hour used by the appliance,
2. volume of room and the standard of ventilation,
3. period of continual use of the room.

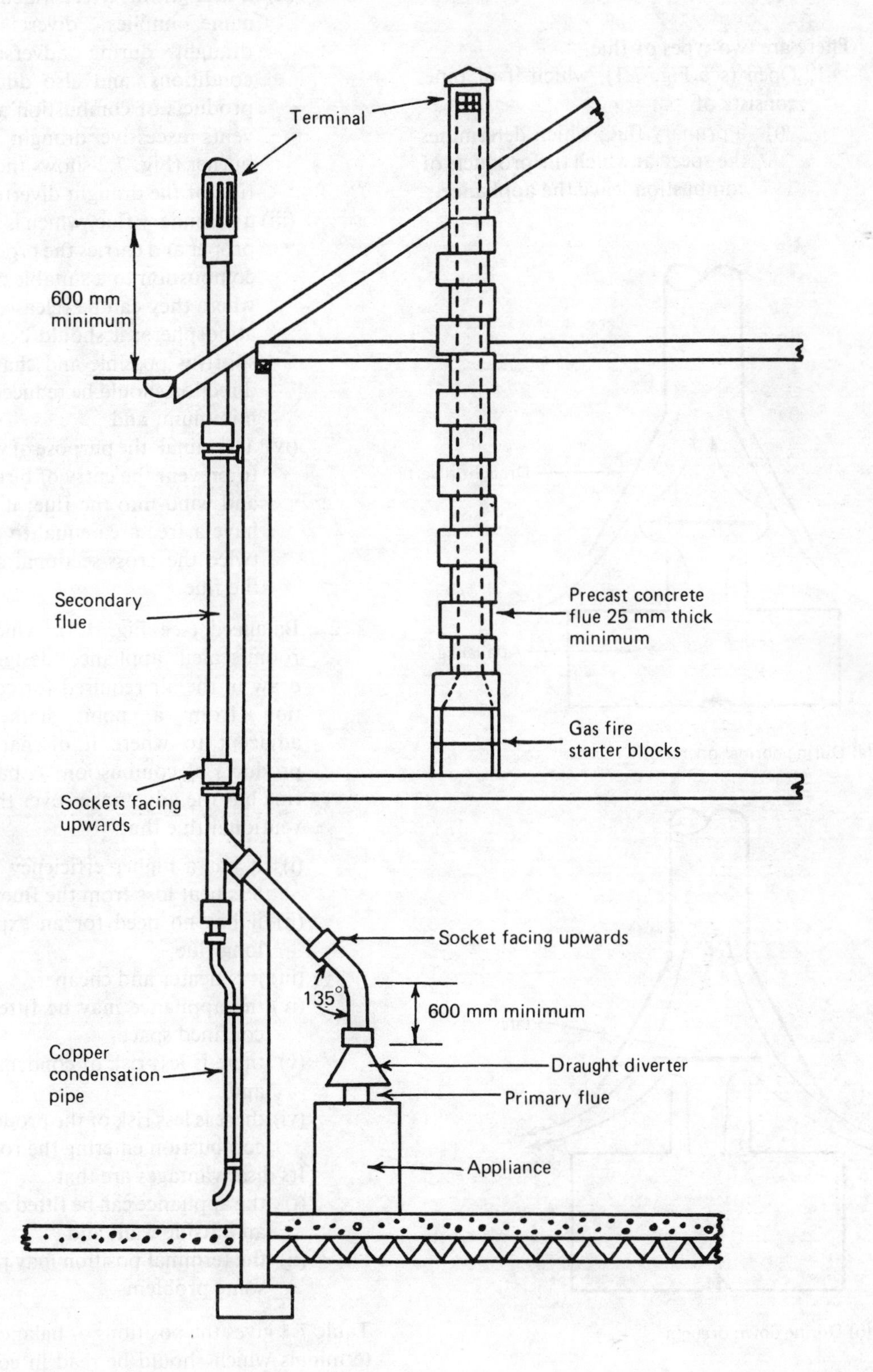

Fig. 7.1 Open gas flues

Types of Gas Flue

There are two types of flue.

1. Open (see Fig. 7.1), which if of pipes consists of
   (i) a primary flue, which determines the speed at which the products of combustion leave the appliance,
   (ii) a draught diverter, which, as the name implies, diverts down-draught during adverse wind conditions, and also dilutes the products of combustion and prevents excessive draught on the burner (Fig. 7.2 shows the operation of the draught diverter),
   (iii) a secondary flue, which is the flue proper and carries the products of combustion to a suitable position where they can be released to the atmosphere: it should be kept as short as possible and changes of direction should be reduced to the minimum, and
   (iv) a terminal, the purpose of which is to prevent the entry of birds, rain and wind into the flue: it should have a free area equal to at least twice the cross-sectional area of the flue.

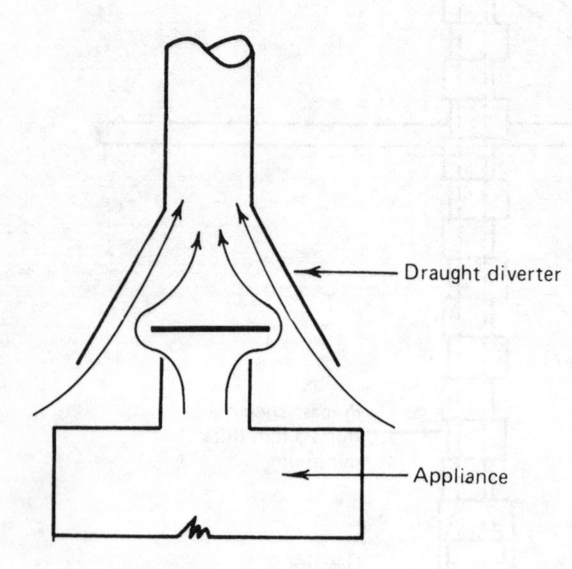

(a) During normal operation

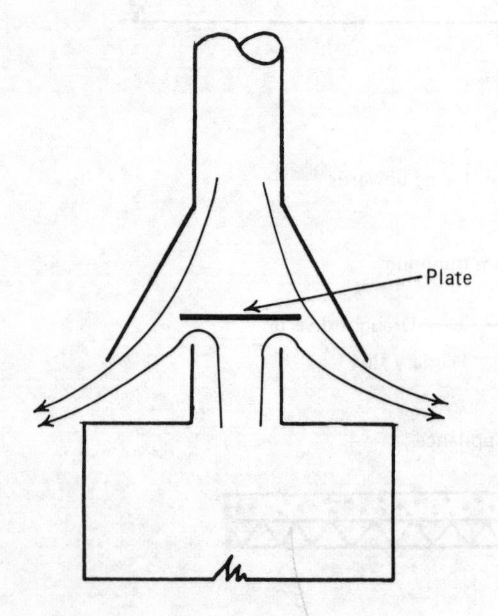

(b) During down draught

Fig. 7.2 Operation of draught diverter.

2. Balanced (see Fig. 7.3), which is a room-sealed appliance designed to draw in the air required for combustion from a point immediately adjacent to where it discharges its products of combustion. A balanced flue has the advantages over the conventional flue that
   (i) it has a higher efficiency due to less heat loss from the flue,
   (ii) it has no need for an expensive long flue,
   (iii) it is neater and cheaper,
   (iv) the appliance may be fitted in a confined space,
   (v) there is less risk of condensation, and
   (vi) there is less risk of the products of combustion entering the room.

   Its disadvantages are that
   (i) the appliance can be fitted only on an outside wall, and
   (ii) the terminal position may present some problems.

Table 7.1 gives the positions of balanced flue terminals which should be read in conjunction with Fig. 7.4.

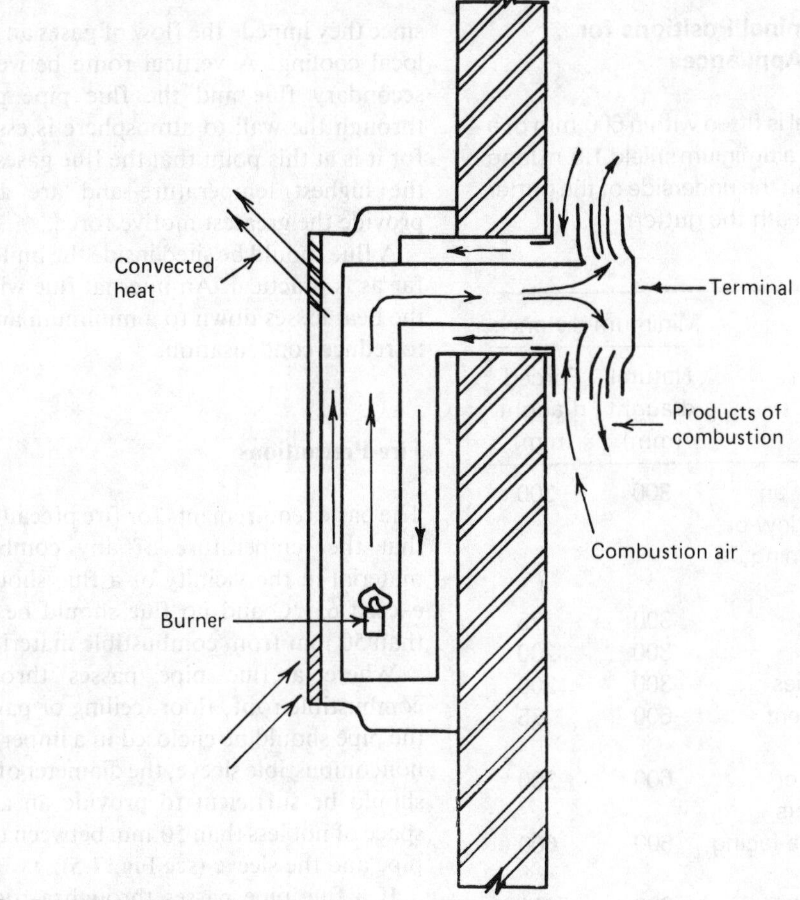

Fig. 7.3 Balanced flue convector heater

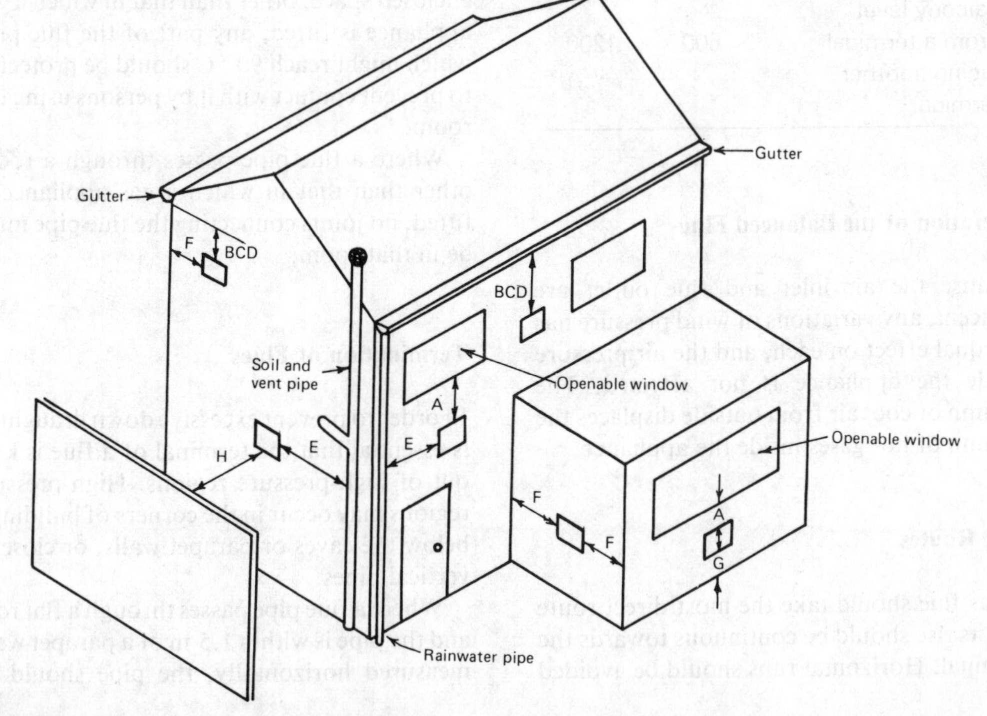

Fig. 7.4 Positions of balanced flue terminals.

Table 7.1 Terminal Positions for Balanced Flue Appliances

Note: if a terminal is fitted within 600 mm of a plastic gutter, an aluminium shield 1.5 m long should be fitted on the underside of the gutter immediately beneath the gutter.

Terminal position	Minimum distance	
	Natural draught (mm)	Forced draught (mm)
A Directly below an openable window or any other opening, e.g. air brick	300	300
B Below gutters	300	75
C Below eaves	300	200
D Below balconies	300	200
E From soil or vent pipes	600	75
F From internal or external corners	600	300
G From a surface facing the terminal	600	600
H Above ground or balcony level	300	300
I From a terminal facing another terminal	600	1200

Operation of the Balanced Flue

Because the air inlet and flue outlet are adjacent, any variations of wind pressure has an equal effect on each, and the air pressure inside the appliance is not affected. The column of cool air from outside displaces the column of hot gases inside the appliance.

Flue Routes

A gas flue should take the most direct route and its rise should be continuous towards the terminal. Horizontal runs should be avoided since they impede the flow of gases and cause local cooling.. A vertical route between the secondary flue and the flue pipe passing through the wall to atmosphere is essential, for it is at this point that the flue gases are at the highest temperature and are able to provide the greatest motive force.

A flue should be sited inside the building as far as is practical. An internal flue will keep the heat losses down to a minimum and help to reduce condensation.

Fire Precautions

The basic requirements for fire precautions is that the temperature of any combustible material in the vicinity of a flue should not exceed 65 °C and no flue should be nearer than 50 mm from combustible material.

Where a flue pipe passes through a combustible roof, floor, ceiling or partition, the pipe should be enclosed in a imperforate, noncombustible sleeve, the diameter of which should be sufficient to provide an annular space of not less than 50 mm between the flue pipe and the sleeve (see Fig. 7.5).

If a flue pipe passes through a room, or enclosed space, other than that in which a gas appliance is fitted, any part of the flue pipe which might reach 90 °C should be protected to prevent contact with it by persons using the room.

Where a flue pipe passes through a room other than that in which a gas appliance is fitted, no joints connecting the flue pipe must be in that room.

Termination of Flues

In order to prevent excessive down draught, it is essential that the terminal of a flue is kept out of high-pressure regions. High-pressure regions may occur in the corners of buildings, below the eaves or parapet walls, or close to vertical pipes.

Where a flue pipe passes through a flat roof and the pipe is within 1.5 m of a parapet wall, measured horizontally, the pipe should be

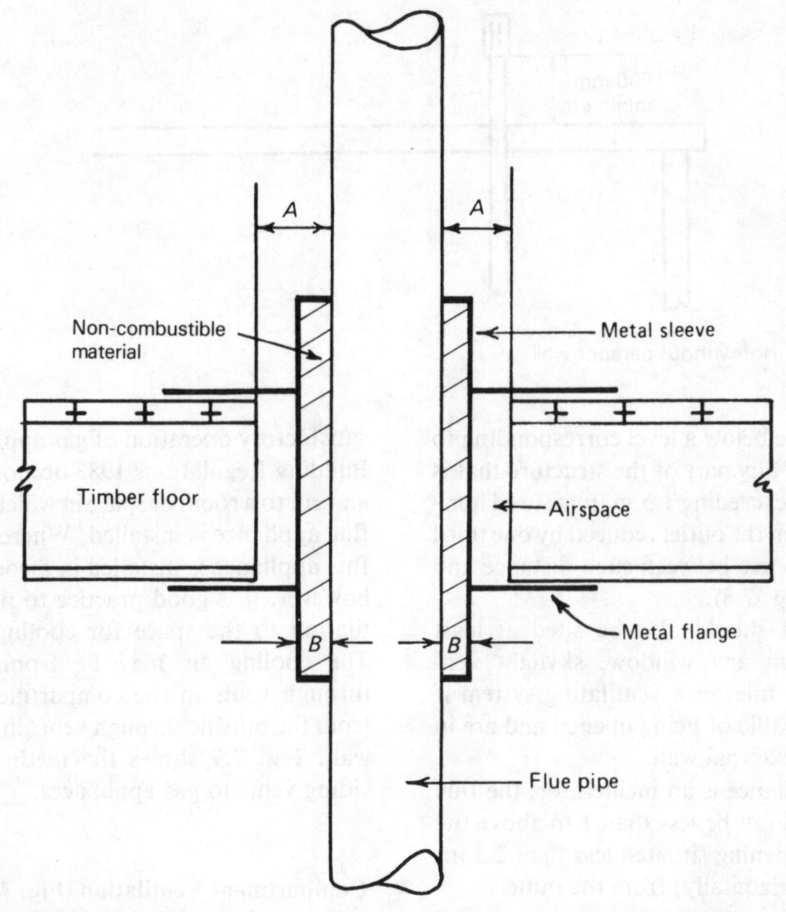

Fig. 7.5 . Flue pipe passing through a timber floor or ceiling. A = 50 mm minimum, B = 25 mm minimum.

carried up at least 600 mm above the top of the wall (see Fig. 7.6).

If a flue pipe passes through a flat roof not having a parapet wall, the pipe should be carried up at least 600 mm above the roof (see Fig. 7.7).

If a flue pipe passes through a flat roof, with or without a parapet wall, the outlet

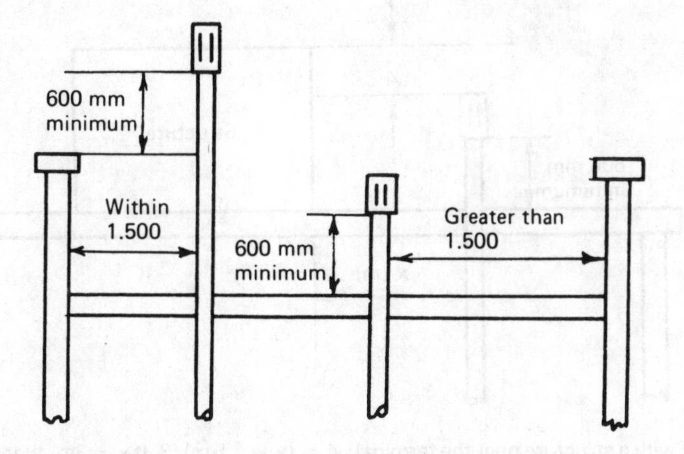

Fig. 7.6 Flat roof with parapet wall.

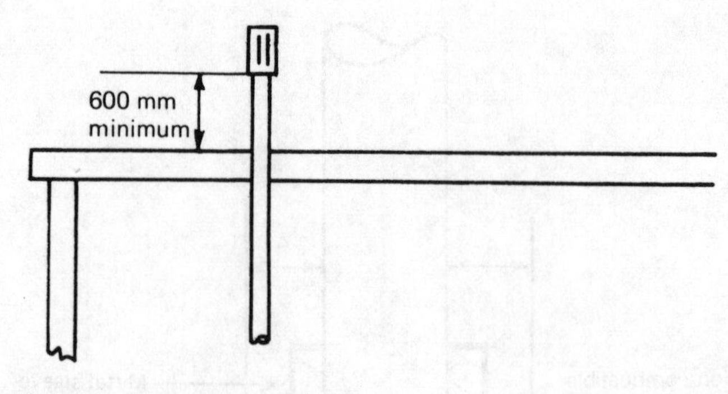

Fig. 7.7 Flat roof without parapet wall.

should not be below a level corresponding to the height of any part of the structure that is at a distance exceeding 1.5 m, measured horizontally, from the outlet reduced by one third of the difference between such distance and 1.5 m (see Fig. 7.8).

The outlet should also be sited at least 600 mm from any window, skylight ventilator or air inlet to a ventilating system if these are capable of being opened and are in any roof or external wall.

If the appliance is an incinerator, the flue outlet should not be less than 1 m above the top of any opening situated less than 2.3 m, measured horizontally, from the outlet.

Ventilation Requirements

An adequate supply of air is essential for the satisfactory operation of gas appliances. The Building Regulations 1985 do not require an air vent to a room or space in which a balanced flue appliance is installed. Where a balanced flue appliance is installed in a compartment, however, it is good practice to provide ventilation to the space for cooling purposes. The cooling air may be from the room through vents in the compartment door or from the outside through vents in the outside wall. Fig. 7.9 shows the methods of providing vents to gas appliances.

Compartment Ventilation (Fig. 7.10)

A boiler compartment is a totally enclosed space within a building, either constructed or modified specifically to accommodate a boiler and its ancillary equipment, excluding

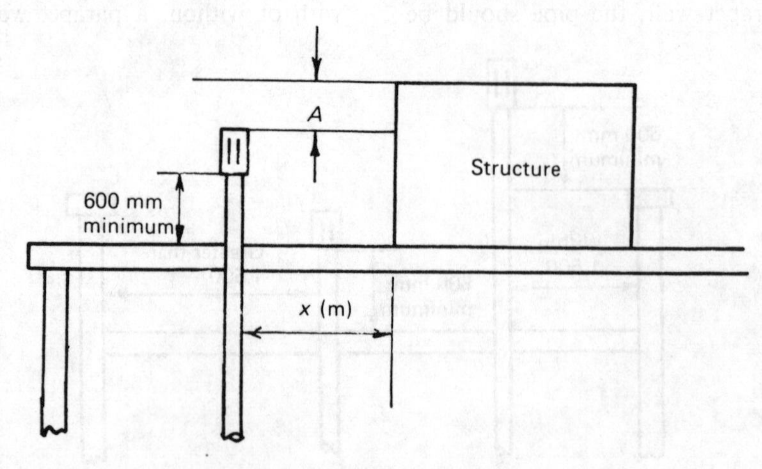

Fig. 7.8 Flat roof with a structure near the terminal. $A = (x - 1.5\,m)/3$. If $x = 3m$, then $A = (3 - 1.5)/3 =$ 500 mm (maximum).

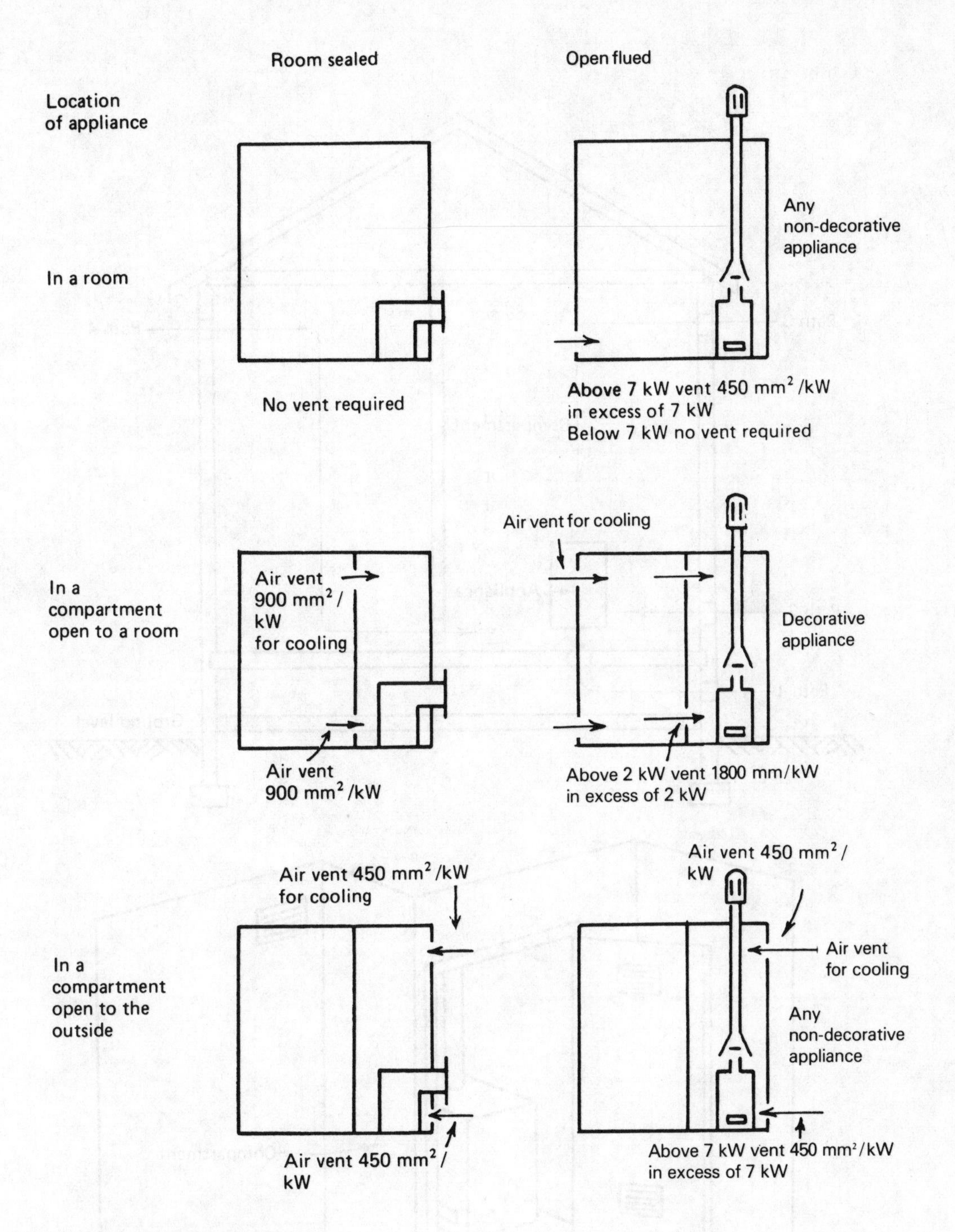

Fig. 7.9 Ventilation of gas appliances.

a gas meter. The requirements for a boiler compartment are as follows.

1. It should be a fixed rigid structure.
2. It should incorporate adequate air vents.
3. It should be fitted with a door of sufficient size to permit the removal of the boiler and ancillary equipment.
4. It should be of sufficient size to permit access for inspection and servicing. It should, however, not be too large or it may be used as a storage cupboard.

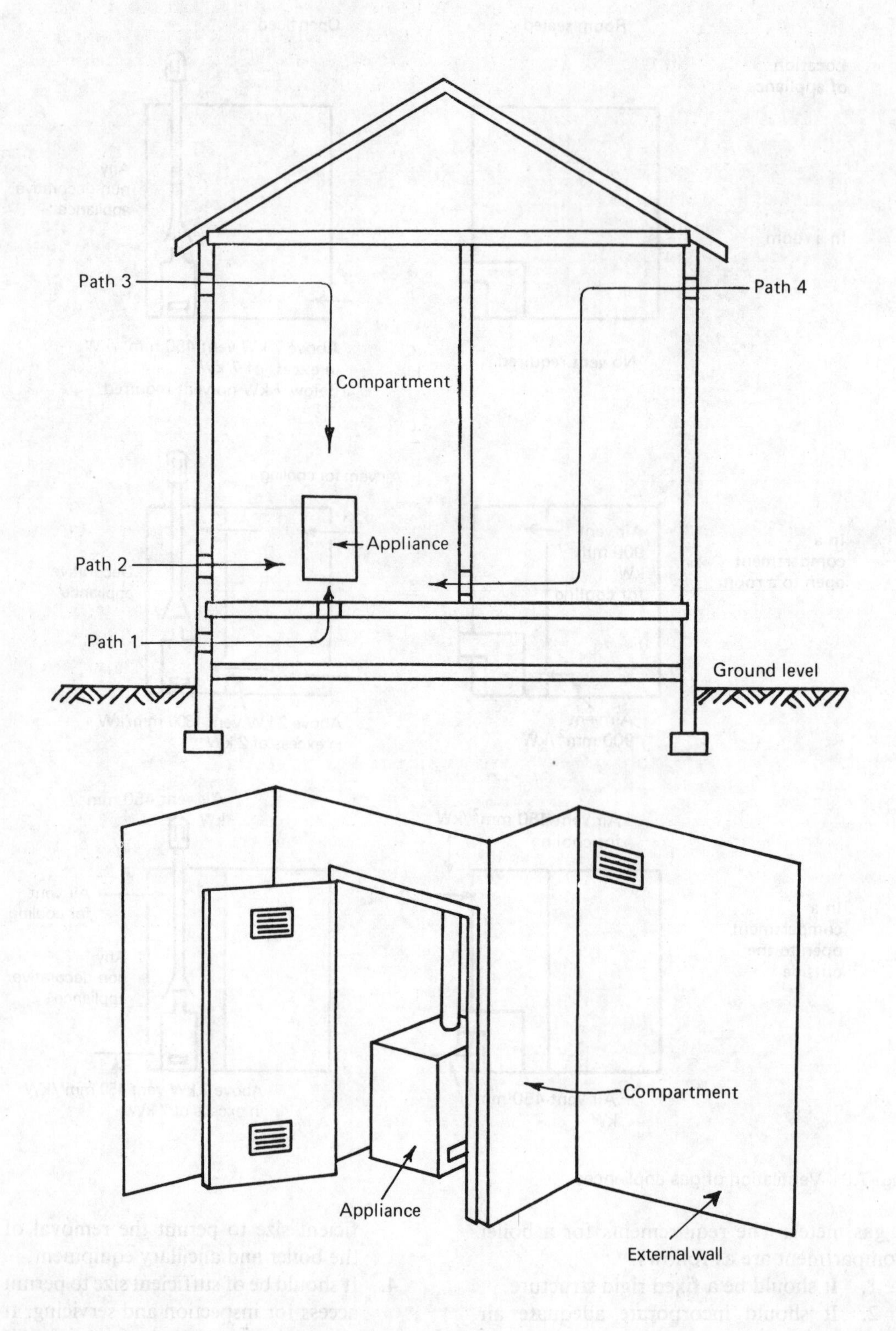

Fig. 7.10 Methods of providing vents to gas appliances fixed inside a compartment.

5. Where the compartment houses an open-flued appliance, neither the door nor the vents should communicate with a bedroom, bed-sitting room or space containing a bath or shower.
6. Any combustible material should be at least 75 mm from any part of the boiler. Alternatively, it should be lined with noncombustible material or treated with a suitable fire-retarding paint.

Table 7.2 gives the minimum effective area requirements for compartment permanent air vents for open-flued and room-sealed appliances. Table 7.3 gives the Building Regulations 1985 minimum unobstructed area of permanent vents to gas appliances.

Table 7.2 Minimum Effective Area Requirements for Compartment Permanent Air Vents (Related to Heat Output)

Permanent air vent position	In a compartment ventilated to a room or internal space (mm^2/kW)	In a compartment ventilated to the outside (mm^2/kW)
(a) *Open-Flued appliances*		
High-level	900	450
Low-level	1800	900
(b) *Room-sealed appliances*		
High-level	900	450
Low-level	900	450

Adventitious Ventilation

Adventitious ventilation is air that can enter a room through cracks around doors and openable windows, etc. Tests have shown that even after weather stripping and double glazing it is virtually impossible to restrict the area of these openings to the room to less than the equivalent of an air vent having an effective area of 3500 mm^2.

An allowance for adventitious openings of 3500 mm^2 can therefore safely by made for open-flued appliances, because the air flow in the flue will draw air into the room through these openings.

Table 7.3 Minimum Unobstructed Area of Permanent Vent

Type of appliance	Volume of room or power output of appliance	Minimum unobstructed area of vent to outside air
Cooker	1. volume of room 6 m^3 minimum, and	
	2. an openable window or other opening to the outside air, and	
	3. vent area for a room volume less than 9 m^3, and	6500 mm^2
	4. vent area for a room volume between 9 m^3 and 11 m^3	3500 mm^2
Open-flued (decorative)	vent area for each kW of rated input over 2 kW	1800 mm^2
Any other open-flued	vent area for each kW of rated input over 7 kW	450 mm^2

Notes: decorative appliances are decorative log and other solid fuel fire effect gas appliances. Extract fans could affect open-flued appliances.

If the room contains two or more appliances, the total air vent requirement is the largest of the following:

1. the total flueless space-heating appliance requirement,
2. the total open-flued space-heating appliance requirement,
3. the largest individual requirement of any other type of appliance.

Location of Air Vent

Where an air vent is fitted in an internal wall, it should not communicate with a bedroom, toilet, bathroom or kitchen. It should be located not more than 450 mm above the floor level, in order to reduce the spread of smoke in the event of a fire. An air vent direct to the outside must not be positioned less than 600 mm away from any part of an open-flue terminal, nor less than 300 mm directly above any part of a balanced flue terminal. An air vent across a cavity wall must include a continuous duct across the cavity. Note: existing or new air vents should be regularly examined to check that they are not obstructed.

Terminals

The outlet of any flue serving a gas appliance must be fitted with a terminal designed to allow free discharge of the products of combustion, minimise downdraught and prevent the entry of any matter that might restrict the flue. Figs. 7.11 and 7.12 show the types of terminal that are fixed to a vertical flue pipe that passes above the roof.

Fig. 7.13 shows the type of terminal that may be used when the flue pipe passes horizontally to the outside. These three different types of terminal are designed to resist the effect of wind pressure, which might cause downdraught in the flue.

Fig. 7.14 shows a stainless steel ridge terminal, and a similar type of terminal may be obtained manufactured from clay. There is normally no high-pressure region at the ridge and therefore this is usually the best possible position for a flue terminal.

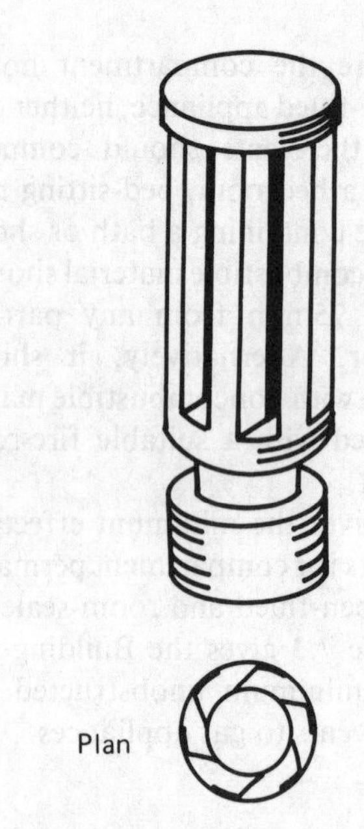

Fig. 7.11 GLC terminal.

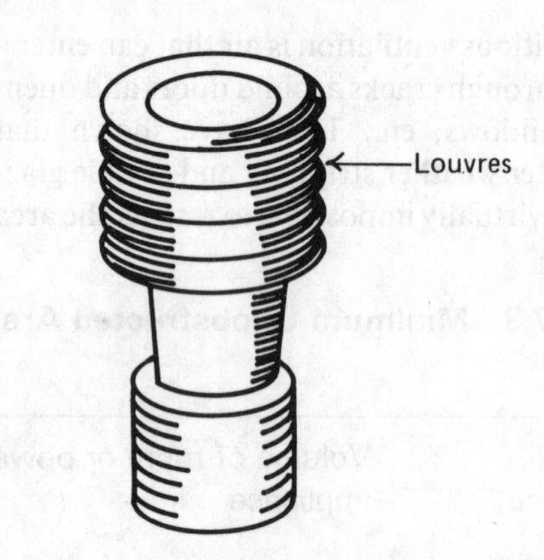

Fig. 7.12 GCI terminal.

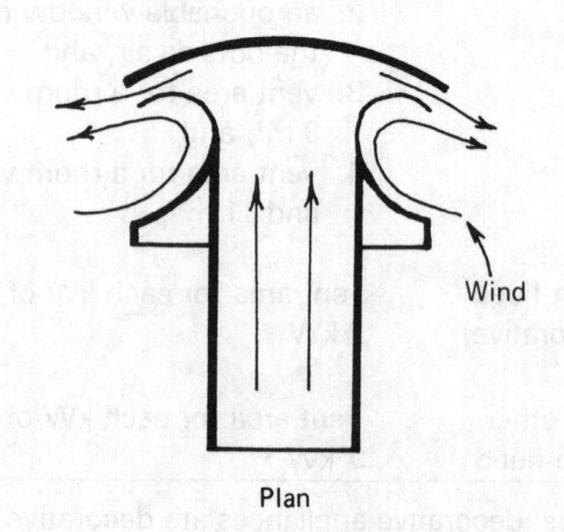

Fig. 7.13 'Denham' terminal for horizontal fitting.

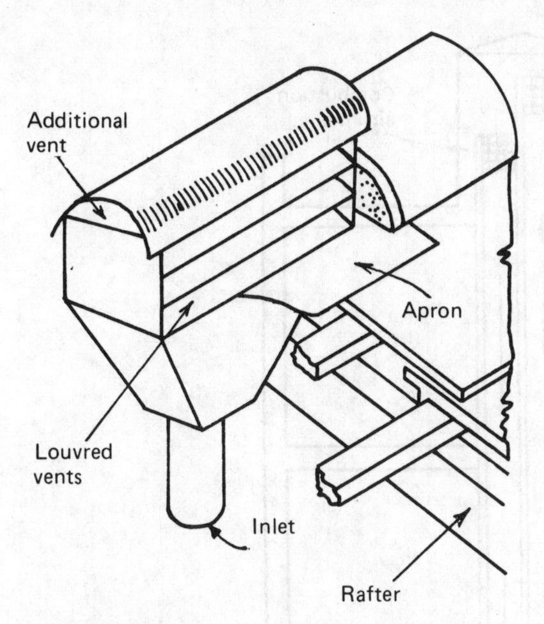

Fig. 7.14 Stainless steel ridge terminal.

Shared Flues

The shared flue system is suitable for high-rise buildings and reduces the cost and space that would be required if separate flues for each appliance were used. There are three systems.

1. Se-duct (Fig. 7.15). Room-sealed appliances may be connected on each side of the flue. The air required for combustion is drawn in from the base of the duct and the entry of this air depends on the type of building structure. If required, a horizontal flue above or below ground level may be used: if the building is built on columns, the horizontal flue is not required.
2. U duct (Fig. 7.16). This type of duct offers an alternative to the Se-duct system in cases where there are difficulties in arranging the air supply at the bottom of the duct. The U duct requires two vertical ducts; the down-flow duct draws air from the roof-level inlet. The room sealed appliances are connected to the upflow duct only.
3. Shunt duct (Fig. 7.17). This type is designed for the flueing of conventional appliances. The system consists

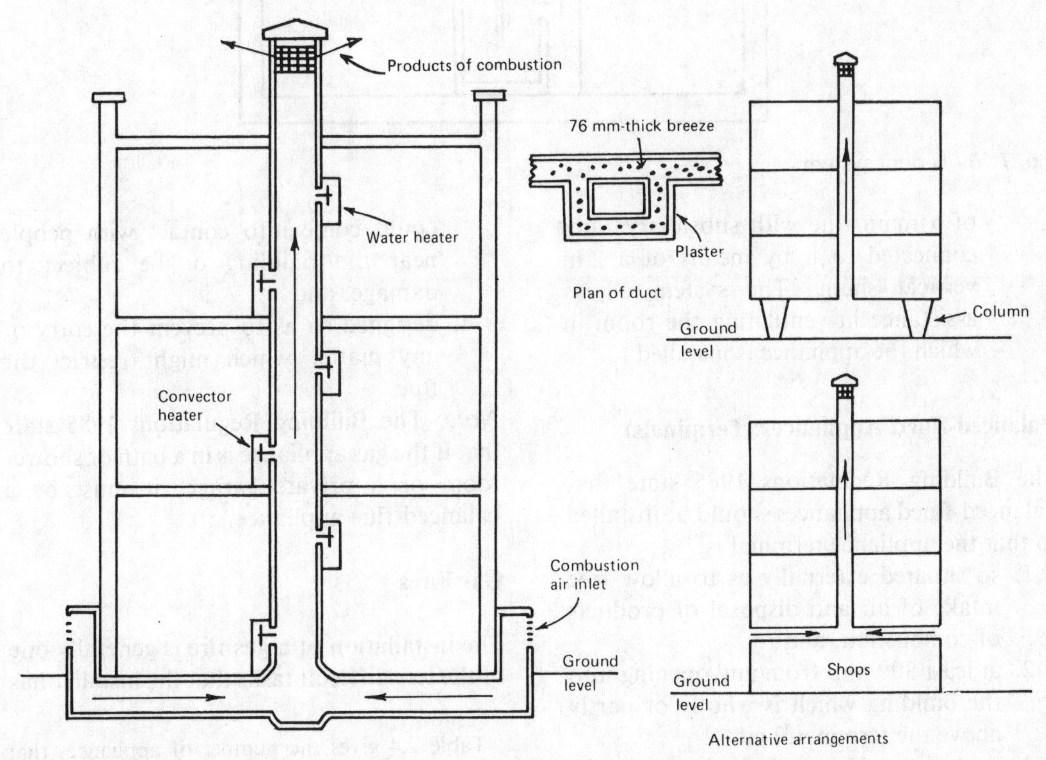

Fig. 7.15 Se-duct system.

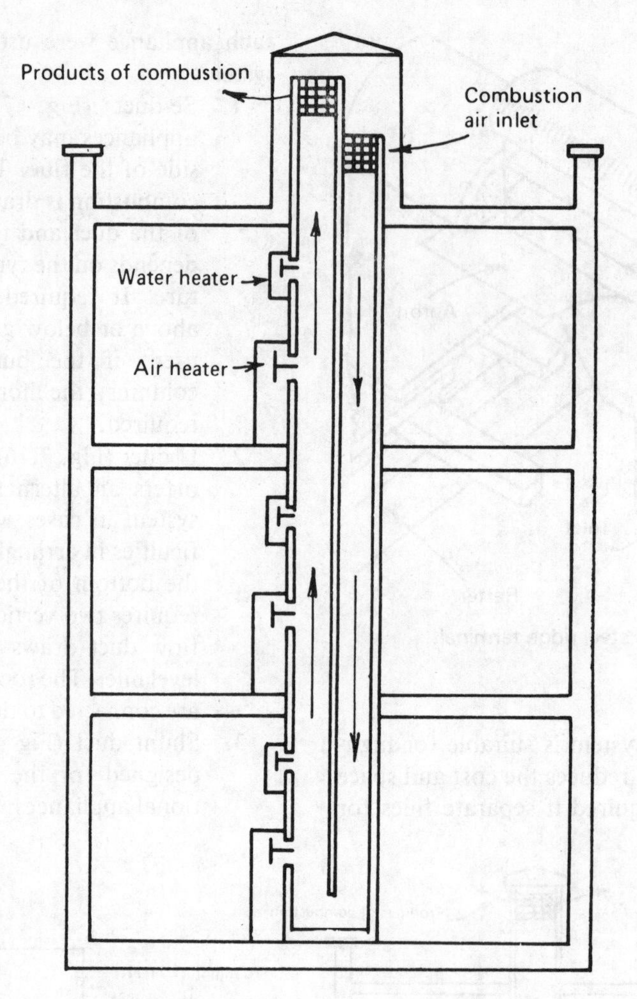

Fig. 7.16 U-duct system.

of a main flue with subsidiary flues connected to it by means of a 2 m vertical shunt. The system allows assistance in ventilating the room in which the appliance is installed.

Balanced-flued Appliances (Terminals)

The Building Regulations 1985 state that balanced-flued appliances should be installed so that the appliance terminal is:

1. so situated externally as to allow free intake of air and disposal of products of combustion, and
2. at least 300 mm from any opening into the building which is wholly or partly above the terminal, and
3. protected with a terminal guard if it could come into contact with people near the building or be subject to damage, and
4. designed so as to prevent the entry of any matter which might restrict the flue.

Note: The Building Regulations 1985 state that if the gas appliance is in a bath or shower room or a private garage, it must be a balanced-flue appliance.

Gas Fires

The installation of a gas fire is generally one of the less difficult tasks that the installer has

Table 7.4 gives the number of appliances that may be connected to flues.

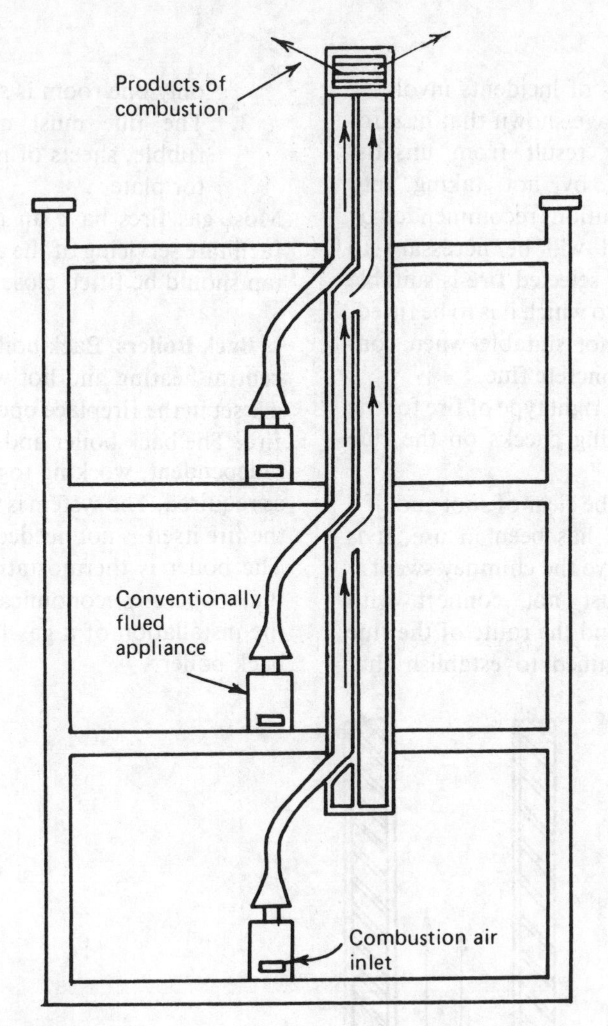

Fig. 7.17 Shunt duct system.

Table 7.4 Appliances Discharging by way of Subsidiary Flues into a main flue

	Nominal cross-sectional area of main flue			
	Not less than 40 000 mm² but less than 62 000 mm²		62 000 mm² or more	
	Maximum number of appliances	Total rating in kW	Maximum number of appliances	Total rating in kW
Convector fire with controlled flue flow having a maximum rate of flow of 70 m^3/h	5	30	7	45
Instantaneous water heater	10	300	10	450
Storage water heater, central heating unit or air heater	10	120	10	180

to carry out. Reports of incidents involving gas fires, however, have shown that hazardous conditions may result from unsatisfactory installation, by not taking into account certain precautions recommended by the manufacturer. It will be necessary to establish whether the selected fire is suitable for the type of flue into which it is to be fitted. Some gas fires are not suitable when connected to a precast concrete flue.

Having selected the right type of fire for the situation, the following checks on the flue should be made.

1. The flue must be clear of soot and, if a solid fuel fire has been in use, it is essential to have the chimney swept.
2. The flue must not connect with another flue and the route of the flue must be examined to establish that only one room is served by the flue.
3. The flue must not be blocked by rubble, sheets of newpaper or restrictor plate.

Most gas fires have an integral tap, but to facilitate servicing of the appliance a separate tap should be fitted close to the appliance.

Back Boilers. Back boilers may be used for central heating and hot water supply. These are set in the fireplace opening behind the gas fire. The back boiler and the fire are entirely independent, working together or separately as required. The system is very economical, as the fire itself is not needed for water heating. The boiler is thermostatically controlled so that it operates economically. Fig. 7.18 shows the installation of a gas fire incorporating a back boiler.

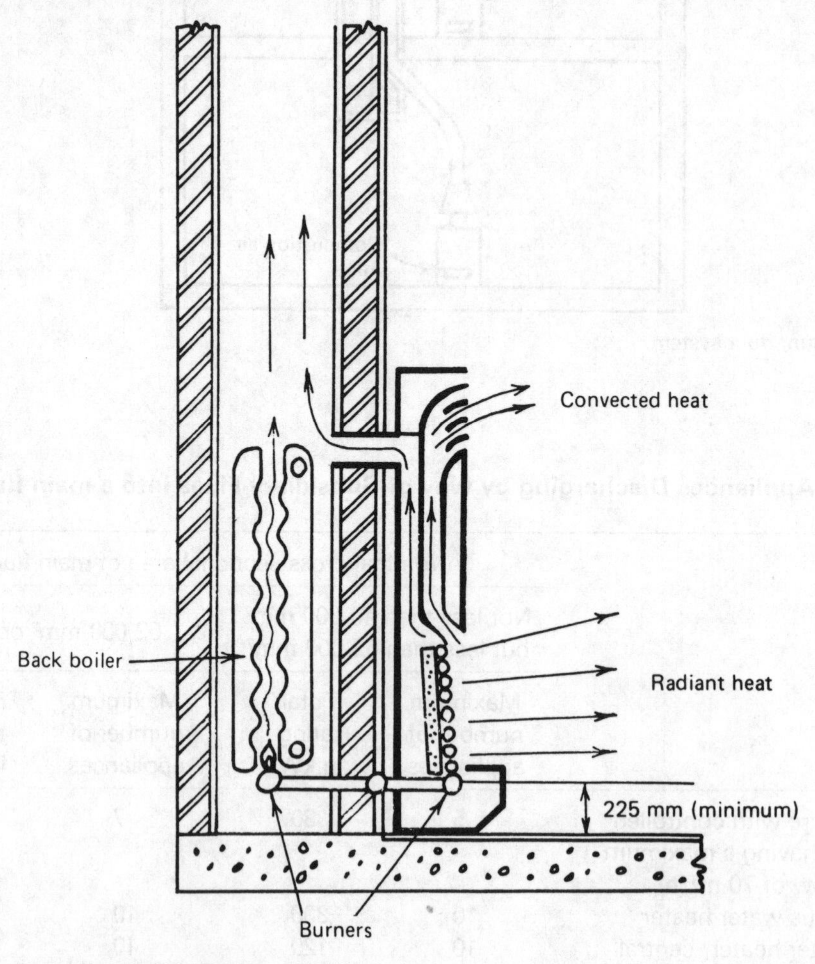

Fig. 7.18 Combined gas fire and back boiler.

The following checks should be made.

1. The vertical height between the gas burner and the finished floor level or floor covering should not be less than 225 mm.
2. The hearth of noncombustible material must extend at least 300 mm from back to front and at least 150 mm sideways beyond the edge of the radiant source.
3. Any part of the rear of the appliance where the temperature is likely to exceed 65 °C, should be separated from any combustible material by an air gap of at least 76 mm.
4. Any combustible side wall less than 500 mm from the radiant heat source must be suitably protected.
5. Any electrical wiring and equipment must comply with the IEE Regulations for the electrical installations in buildings and also with the requirements of CP 321.
6. The gas pressure should be checked and adjusted with the manufacturer's instructions.

Exercises

1. Define
   (a) appliance ventilating duct,
   (b) room-sealed appliance,
   (c) subsidiary flue,
   (d) primary and secondary flues.
2. State the types of material that may be used for gas flues and compare the advantages of each.
3. State the purposes of flues and the factors governing their design.
4. Sketch a conventional type of gas flue for a gas water heater showing all the components.
5. Sketch a section through a draught diverter and explain its operation.
6. Sketch a section through a balanced flue appliance, explain its operation and state its advantages over a conventional type of flued appliance.
7. State the considerations that must be taken into account for
   (a) routes for flues,
   (b) fire precautions for flues.
8. State the positions required for flue terminals and explain the reasons for using these positions.
9. Explain the requirements for the ventilation of gas appliances including 'compartment ventilation'.
10. Sketch a flue terminal for a position
    (a) where a flue pipe passes through a roof,
    (b) where a flue pipe passes to a ridge,
    (c) where a flue pipe passes through a wall.
11. Sketch vertical sections and explain the operation of the following types of shared flue:
    (a) Se-duct,
    (b) U duct,
    (c) shunt duct.
12. State the requirements of the Building Regulations 1985 for the connections of room-sealed appliances to shared flues.
13. Sketch and describe the method of installing a gas fire and state the checks that must be made before and after installing the fire.

Chapter 8
Oil Firing, Calorific Values

Oil Burners

Because oil fuel in bulk will not burn easily, it is necessary to provide some means by which small quantities can be vapourised or atomised and mixed with air in the correct ratio to form a combustible mixture, which can then be ignited and burned to release heat. There are two types of oil burner: vaporising and atomising.

Vaporising Oil Burners

These are by far the most commonly used type in domestic heaters and are almost silent in operation. They can be installed in a kitchen or living room. There are three classes of vaporising burner: (a) rotary vaporising burner, (b) natural draught pot vaporising burner and (c) forced draught pot burner.

Rotary vaporising burners. These are fully automatic burners suitable for heaters providing 12–44 kW. The burner is entirely dependent on a mains electrical supply and a flue capable of providing a steady draught of about 5 Pa. There are two forms of this type of burner, namely Wallflame and Dynaflame burners.

Wallflame burners (Fig. 8.1) consist of a steel plate securing a centrally placed electric motor. The armature of this motor is wound on a hollow metal shroud and is termed the rotor. The bottom of the shroud dips into an oil well, which also supports the trust bearings of the rotor spindle.

A constant-level oil control feeds oil into the well and maintains it at a level just covering the edge of the shroud. The shroud is circular, with its internal diameter increasing towards the top, from which two holes connect with a pair of oil distributor pipes, which are inclined outwards and project through the base of the plate.

Above these tubes and mounted on the rotor spindle is an air fan, and on top of the steel plate there is a ciccular refractory hearth. Within the circumference of the hearth there is a sheet metal band arranged so as to provide an air gap to insulate the lower part of the boiler from the burner flame. Fixed on the hearth, concentric with the backing band, is a trough-shaped flame rim on which is mounted a number of grilles.

A high-tension electrode is fitted into the base of the hearth and projects into the flame rim trough. On some models of this type of burner, flame rim heating is also provided to assist in flame ignition.

The burner is controlled in accordance with demands for heat and therefore operates on the on/off sequence. As soon as the burner is switched on, a high-tension electric arc is established between the electrode and the flame rim. On models incorporating a rim heater, the start-up sequence begins with this heater being energised. The motor drives the rotor clockwise and oil is lifted through the shroud and oil distribution tubes by cen-

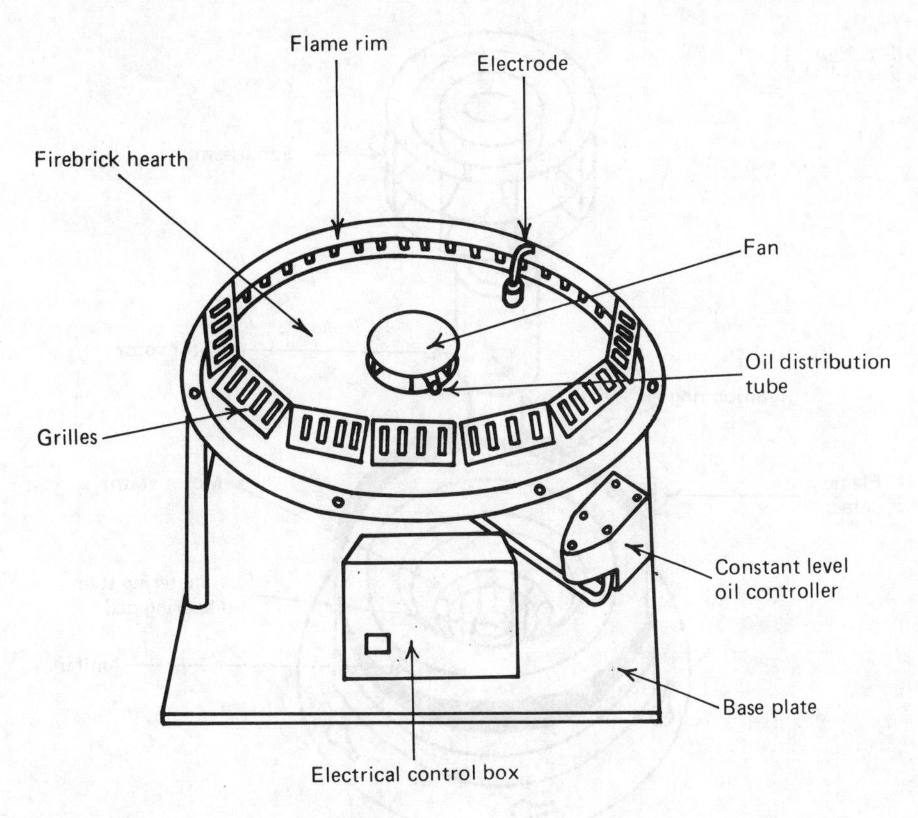

Fig. 8.1 Wallflame rotary vaporising burner.

trifugal force and ejected as a horizontal spray onto the flame rim. The droplets of oil that fall on the hot spot on the rim, created by the high-tension spark, vaporise rapidly and are ignited by the spark.

Heat is radiated from the burning vapour and flame is quickly established around the complete circumference of the hearth. The grilles on top of the flame rim shortly become red hot and radiate heat down onto the oil to cause it to vaporise almost instantaneously. When the burner is firing correctly, the flame burns through the grilles providing a wall of flame around the circumference, hence the name 'wallflame'.

Air for combustion is drawn through the hollow air casing of the motor and is projected horizontally across the hearth by the fan. An air shutter provides for regulation of the volume of combustion air as required.

Dynaflame burners (Fig. 8.2) have many unique features. Exclusive to this burner are a constant-output fuel-metering system, which ensures that the correct amount of fuel is always available, and a controlled recirculation of gases resulting in a blue flame combustion. Operation is controlled from a special control box and ignition power pack.

The burner consists of two main parts: (a) the casing assembly and (b) the rotary assembly. The casing assembly is attached to the boiler by engaging the slotted holes into fixing bolts in the boiler base plate. The assembly contains the stator of the high-speed burner motor, a central fuel reservoir and an adjustable combustion air shutter.

The mounting flange contains a fixed airway to allow air to flow into the combustion chamber and also houses an ignition plug, ignition ring and flame detector. A vertical fuel-metering column and external helical grooving is mounted within the base of the fuel reservoir, and this also provides a bearing for the rotating assembly.

The rotating assembly comprises a central tube, with internal sleeve and bearing, and a

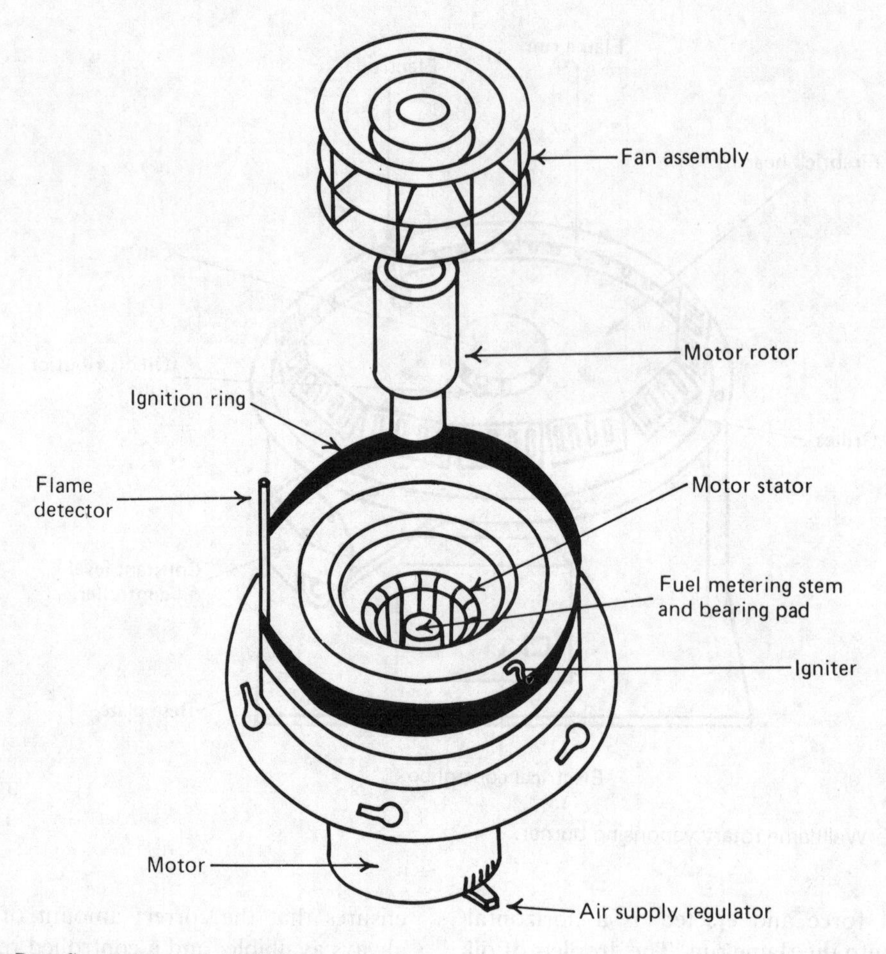

Fig. 8.2 Dynaflame rotary vaporising burner.

double-tiered fan assembly attached to its upper end by means of an internal cone. The rotor of the burner motor is attached around the tube.

A flame stabiliser, which is fitted around the base of the combustion chamber in the boiler and above the burner, assists in achieving a stable flame pattern. The oil is supplied to the burner via a constant-level oil control device.

The boiler thermostat controls the start and stop operation of the burner, unless this is overidden by a time control, programmer or frost thermostat.

When the electrical supply is switched on, a minimum of 95 s elapses before the thermostat can start the burner. When the thermostat initiates the starting cycle, the igniter is connected to the electrical supply and reaches ignition temperature within 25 s. A change-over switch in the ignition circuit then operates to start the burner motor and the rotating assembly runs at full speed.

A metered quantity of oil is lifted up a spiral passageway between the rotating tube and the fuel-metering column and passes inward and upwards through holes at the top of the metering column, entering the base of a hollow fuel-discharge cone. The rotating effect lifts the fuel uniformly to the uppermost edge of the cone and discharges it from the edge of the upper flange, in the form of fine particles, into the blades of the surrounding top fan assembly.

Air is drawn through the fan from the combustion chamber, mixes with the oil and this fuel passes over the hot igniter. Local ignition occurs immediately and a full circle of flame is quickly established in the ignition ring, which surrounds the fan. Combustion air supplied by the lower fan assembly enters the combustion chamber through an airway in the mounting flange. Flame is established around the outer wall of the combustion chamber and the fan circulates the products of combustion, quickly establishing a blue flame condition.

The flame detector quickly senses the presence of flame and causes the igniter to be switched off. When the boiler thermostat senses the preset water temperature, the burner motor is switched off and a minimum delay of 96 s must elapse before the thermostat can again initiate the starting cycle.

Natural Draught Pot Vaporising Burner (Fig. 8.3). This is the simplest type of burner and is commonly fitted in heaters with an output in the range 6–18 kW. They are designed to serve central heating boilers or air heaters. The heaters must be connected to a flue capable of creating a draught of 10–20 Pa, depending on the type of heater and its heat output. The burner is almost silent in operation and may be installed in a kitchen or livingroom.

The heat output from the burner may be sufficient to cause high water temperatures and a 'heat leak' in the form of a bathroom radiator or towel rail is frequently recommended.

The burner consists basically of a circular container or pot, which in some cases has a smaller-diameter opening at the top with rows of holes widely separated at the bottom, and close together at the top, to admit the required volume of air for combustion. The holes at the top are sometimes formed so as to cause a swirl to the incoming air, which assists the mixing of the oil vapour and air.

Air is admitted progressively through the holes at the side of the cylinder, so that as the firing rate is increased the flame has to rise up the pot to obtain sufficient air for combustion. Oil is metered at a low flow rate from a constant level controller and spreads a thin film of oil at the base of the pot.

Heat generated by this burning thin film of oil causes vaporisation, and oil vapour rises up the pot but is too thick to ignite and requires more oxygen. When this oil vapour comes into contact with the air entering through the lowest row of holes, it mixes with it and ignites. The flame then increases in size, radiates more heat and the vaporisation

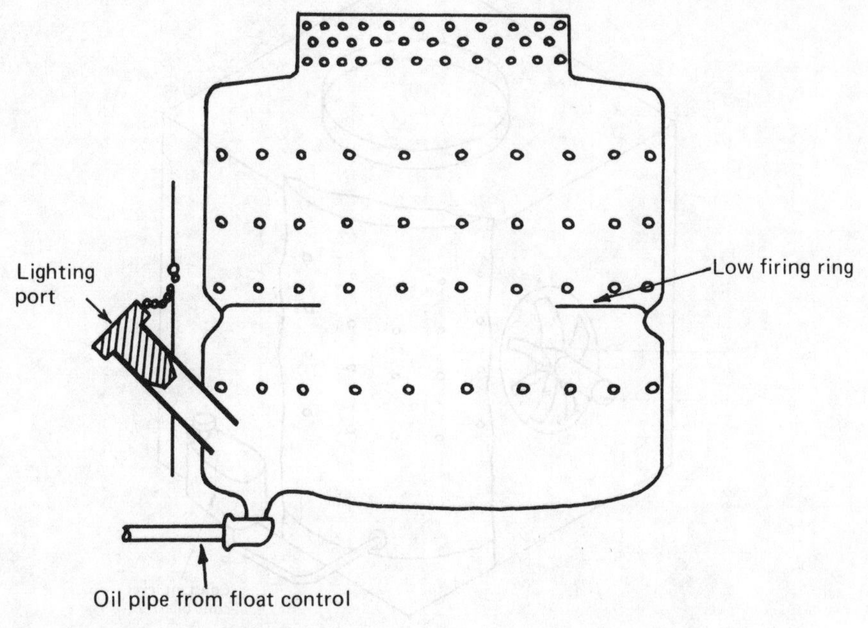

Fig. 8.3 Natural draught oil burner.

process continues. Oil vapour continues to rise and when it obtains sufficient air it ignites until the amount of air mixing with the vapour results in the flame burning out at the top of the pot, as shown in Fig. 8.4.

Note: all appliances equipped with natural draught burners require a draught stabiliser to ensure that a steady draught is maintained at the burner.

The gradual introduction of air is very important, as too much air at the base of the pot would cause the flame to be too close to the oil film, overheating it and causing the formation of carbon deposits.

The high-firing ring at the top of the pot is an important part of the design and forms a throat which promotes the final mixing of air and oil vapour. The burner also contains a low-fire ring above the first row of air holes, but this will not normally allow a low flame below one-third of the high firing rate.

As mentioned previously, overheating of the hot water supply may occur when the central heating is switched off and some form of 'heat leak' is therefore sometimes required.

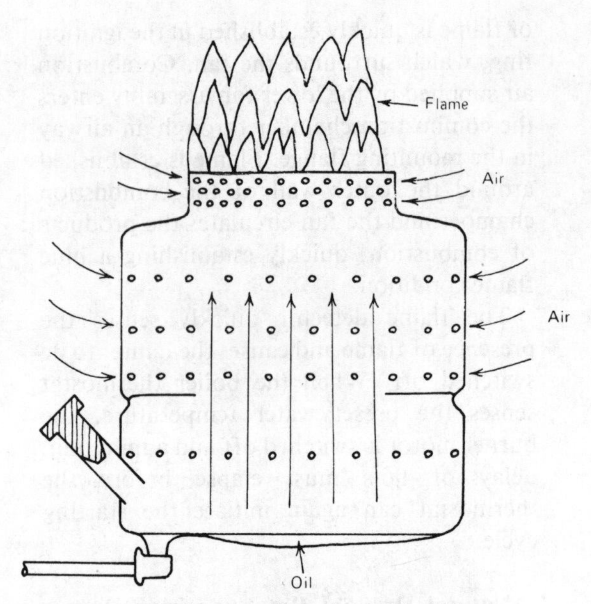

Fig. 8.4 Natural draught vaporising burner operating on high flame firing.

Forced Draught Pot Burner (Fig. 8.5). The design of this burner is similar to that of the natural draught pot burner, except that the pot is enclosed inside a casing into which air required for combustion is supplied by a fan.

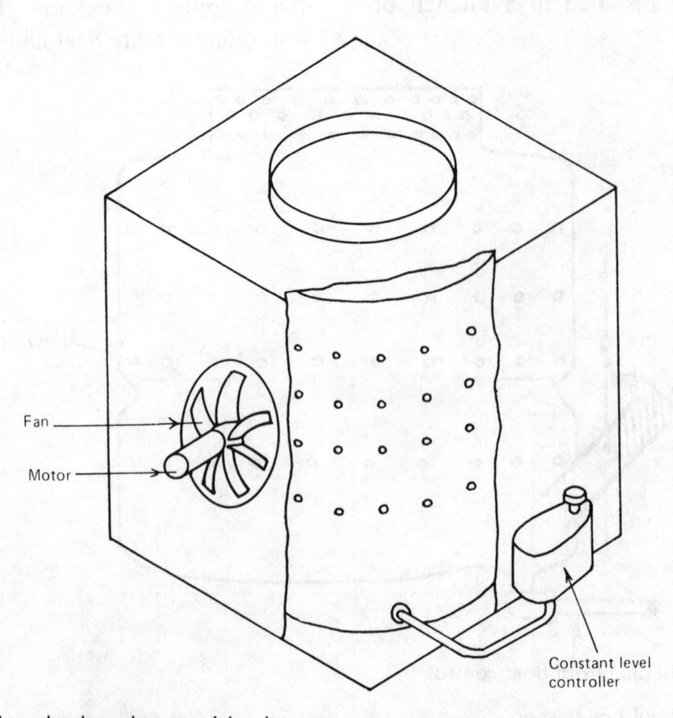

Fig. 8.5 Forced draught draught vaporising burner.

Oil Metering and Constant Level Control. Because the oil feed to vaporising burners requires to maintain a constant level of oil, some form of control is required. The function of the control is to maintain a constant level of oil at the burner, irrespective of the level of oil in the storage tank, and to meter the oil supply to the burner at the correct rate.

Fig 8.6 shows a section of a double-float constant-head controller which also has a safety function in providing a secondary cut-off should the inlet needle valve continue to pass oil when the correct operating level is exceeded. The controller operates as follows.

1. Oil under pressure from the storage tank flows through the filter and inlet needle valve.
2. The oil is maintained at a constant level by means of float A which raises or lowers the inlet valve as required.
3. When the needle valve makes contact with its seating, it shuts off the supply of oil from the storage tank.
4. The inlet valve normally takes up a position such that the inflow of oil to the burner is sufficient for the heat required.
5. If the needle valve fails to close, because of a faulty seating or dirt accumulating between the valve and the seating, the oil level rises and flows over the wier into the trip chamber.
6. Float B then rises and actuates the trip mechanism and the spring forces down lever C onto the needle valve. The needle valve is thus forced down against its seating with sufficient force to dislodge any particles of dirt.

Atomising Oil Burners (Fig. 8.7)

Atomisation is an alternative to vaporising the oil before mixing it with air to cause it to burn. Although there are many types of atomising burner, the pressure-jet type is the most common. This is chiefly because of its compact design and relatively silent operation. They are available with outputs of from 15 kW upwards.

The burner operates as follows.

1. Oil is pumped to the atomiser, which breaks up the oil (atomisation) into a very fine spray.
2. The fan forces a controlled amount of air through the air director, so that it mixes with the oil spray. The air director maintains a consistent air pattern, stabilises the flame and creates the required flame shape.

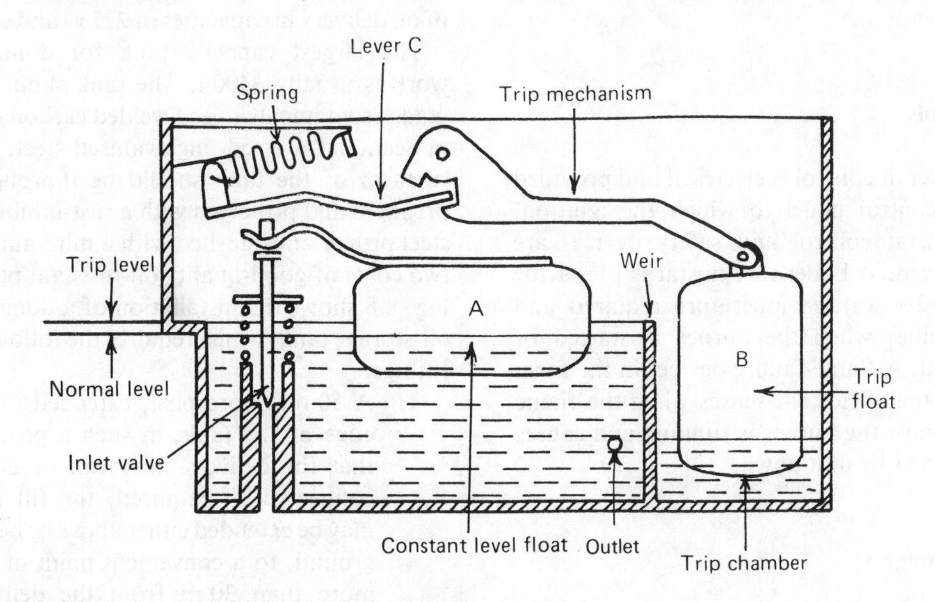

Fig. 8.6 Double-float constant-level oil controller.

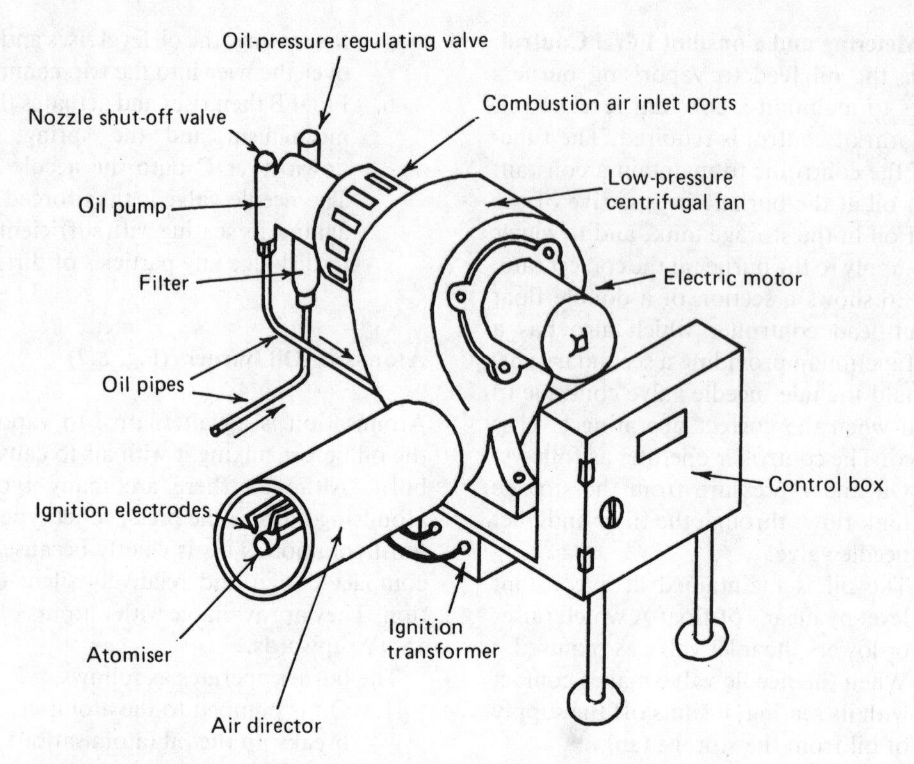

Fig. 8.7 Automatic pressure jet burner.

3. The oil mixture is ignited by the ignition electrodes, which require 10 kV across the electrodes. The burner then operates in an on/off sequence in accordance with the demands for heat.

Controls

The overall control is electrical and provided by a control panel to which the ignition, thermostat control and safety devices are connected. A boiler thermostat is preset for the boiler water temperature required and determines when the burner is started or stopped. A flame-failure device, in the form of photoelectric cell, senses when the flame fails whilst the burner is running and causes the burner to shut down.

Oil Storage

Generally, a rectangular tank with a top designed to shed water and with a working capacity of 1136 l is satisfactory for installations of up to about 13 kW. It is, however, in the client's interest to install a larger tank so that advantage may be taken of the lower cost of oil delivery in capacities of 2275 l and over.

The largest capacity tank for domestic work is usually 3400 l. The tank should be made from ungalvanised welded carbon steel or sectional pressed ungalvanised steel. The surfaces of the tank should be thoroughly prepared and protected with a rust-inhibiting steel primer and finished with a minimum of two coats of good-quality oil-resistant paint. Fig. 8.8 shows the installation of a domestic oil-storage tank, which requires the following fittings.

1. A 50 mm bore pipe, extended to the edge of the tank, in such a position that the delivery hose can be easily connected. If required, the fill pipe may be extended either above or below ground, to a convenient point of not more than 30 m from the delivery tanker. A buried extension line should

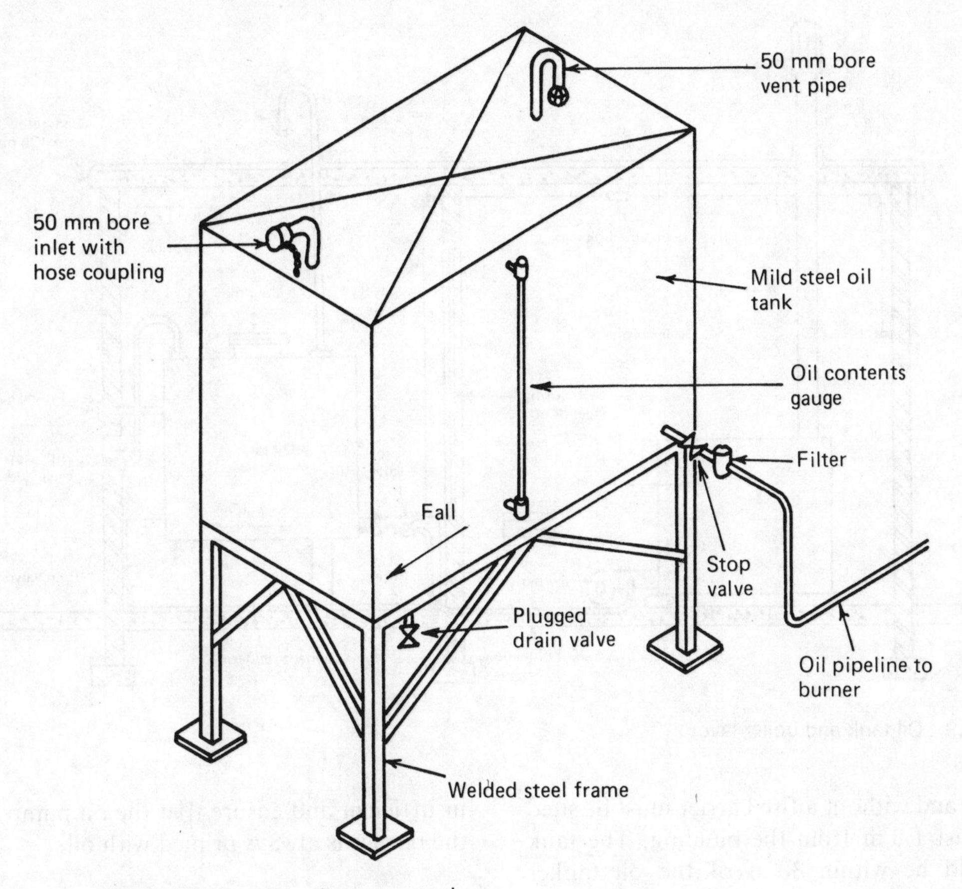

Fig. 8.8 Installation of domestic oil storage tank.

be terminated within a manhole, or the inlet connection must be raised to about 300 mm above ground.

2. A vent pipe must always be fitted and its bore must always be at least equal to that of the fill pipe. It should rise above the tank no more than 600 mm, unless the tank is installed inside a tank room. The open end of the pipe must be fitted with an open-mesh wire balloon. Overflow warning devices must never be fitted to the vent pipe, since with certain types internal tank pressures can rise when the tank is being filled and damage can result.
3. An isolating valve must always be fitted in the tank outlet, in an accessible position for connecting the oil supply. It is important that the tank outlet tapping is slightly above the bottom of the tank to ensure that no water or sediment can be drawn into the burner. Other fittings include a contents gauge and a plugged drain valve.

Location of Oil Tank

The tank should, if possible, be sited above ground and, for domestic work, in the open. Larger tanks are usually sited inside an oil tank room adjacent to the boiler. Fig. 8.9 shows an oil tank and boiler layout. The wall and self-closing steel door between the tank room and boiler room must have a minimum fire resistance of 4 h for tanks over 4000 l, 2 h for tanks up to 3400 l and 1 h for tanks up to 1250 l.

A tank inside a garage must be separated from the vehicle space by a fire-resisting wall and ventilated to the open air. A tank in the

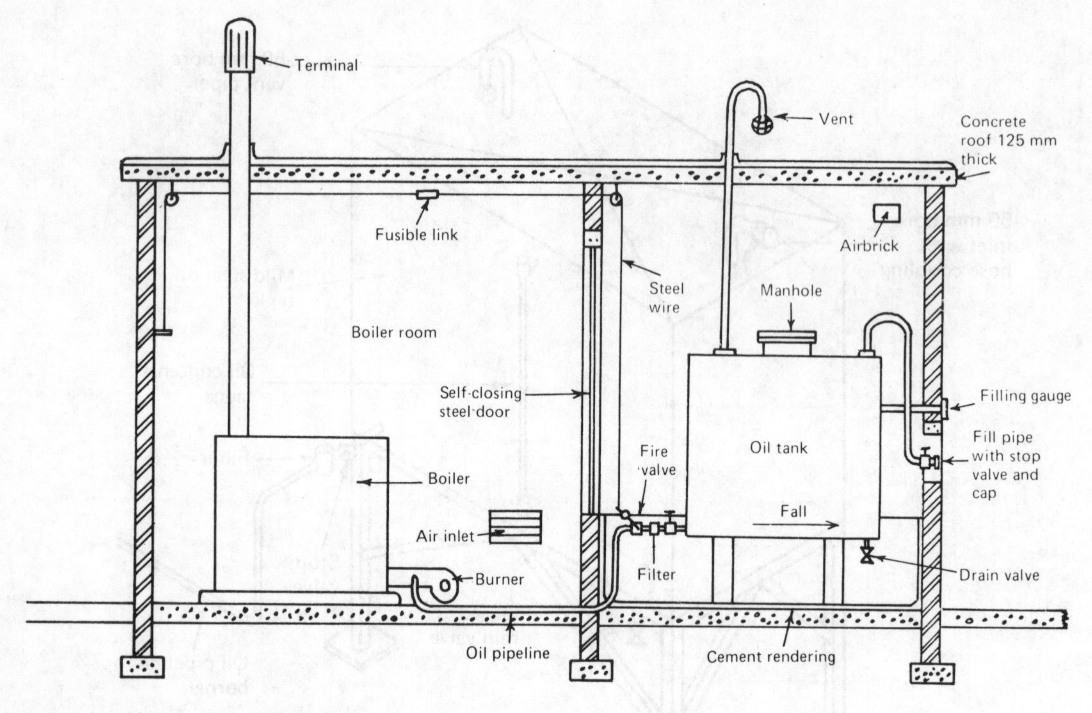

Fig. 8.9 Oil tank and boiler layout.

open and without a fire barrier must be sited at least 1.8 m from the building. The tank should be within 30 m of the oil tanker vehicle access point, unless an extended fill line is provided. The vehicle must not go over the customer's property.

Oil Supply Systems

Annealed copper pipe is excellent for oil supply lines and enables the line to be run with the minimum of joints. Fig. 8.10 a shows the supply to a vaporising burner. These normally require a one-pipe feed system and depend for their supply on an adequate head of oil in the tank. Fig. 8.10 b shows the supply to an atomising burner with a one-pipe feed. The head of oil from the outlet pipe to the centre of the burner must be at least 300 mm. This amount of head will ensure that the oil pump on the burner is always primed. The maximum head above the burner is 3 m. Fig. 8.10 c shows the supply to an atomising burner with a two-pipe feed. This is required when the head of oil above the burner is insufficient and ensure that the oil pump on the burner is always primed with oil.

Rooftop Boiler Room (Fig. 8.11)

A rooftop boiler room reduces the length of flue pipe to a minimum, saves space and ensures adequate air for combustion. The oil requires to be pumped up to a service tank sited above the boiler burner. The pump is switched on automatically by an oil-level swith fitted to the service tank. An overflow pipe is required on the service tank to prevent overfilling.

Fire Valves

A fire valve is required to shut off the supply of oil to the burner automatically in the case of a fire. There are two main types.

1. A mechanical type, comprising a spring-loaded or free-fall weight-operated gate valve held open by a ten-

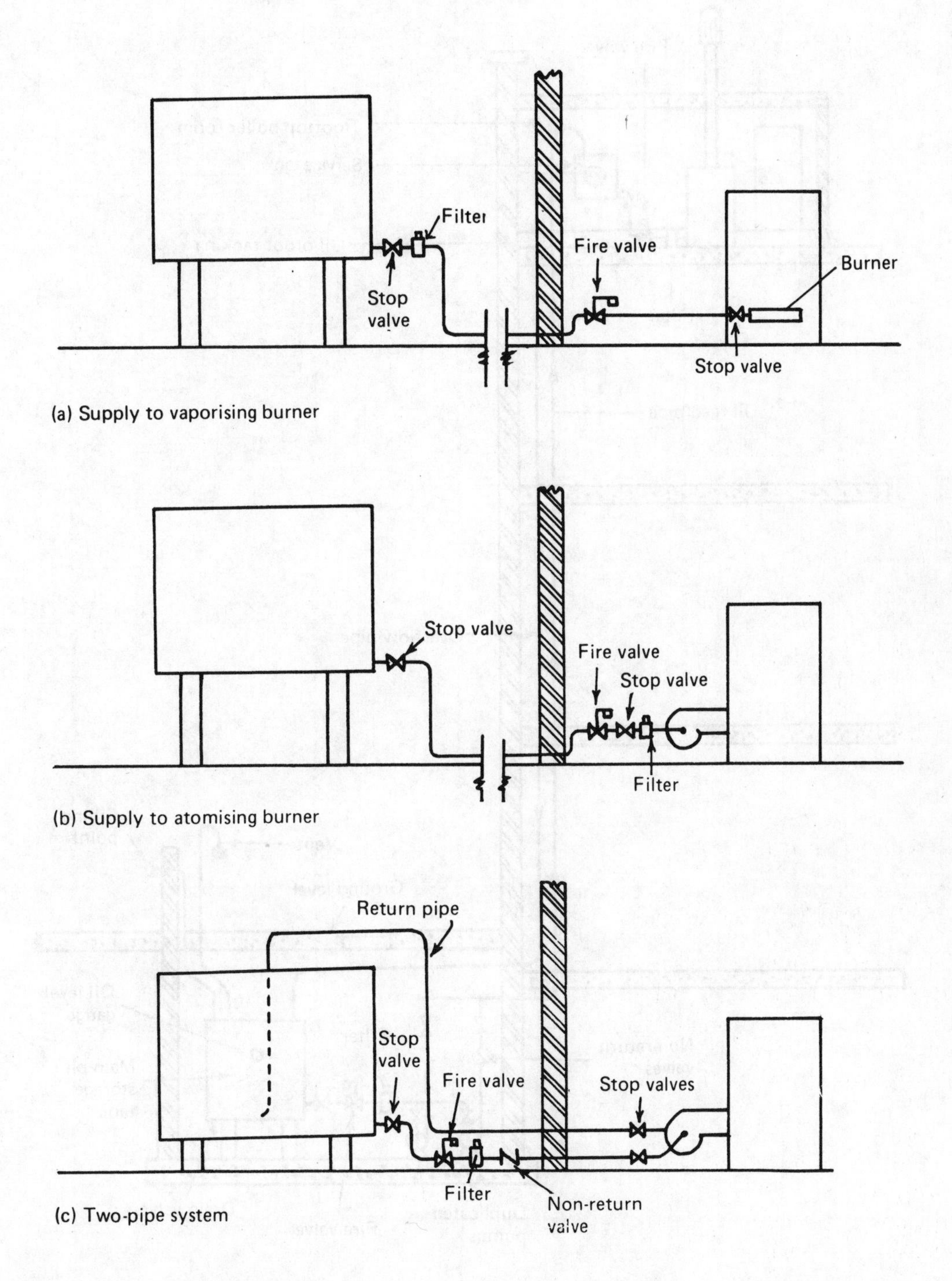

Fig. 8.10 Oil supply to burners.

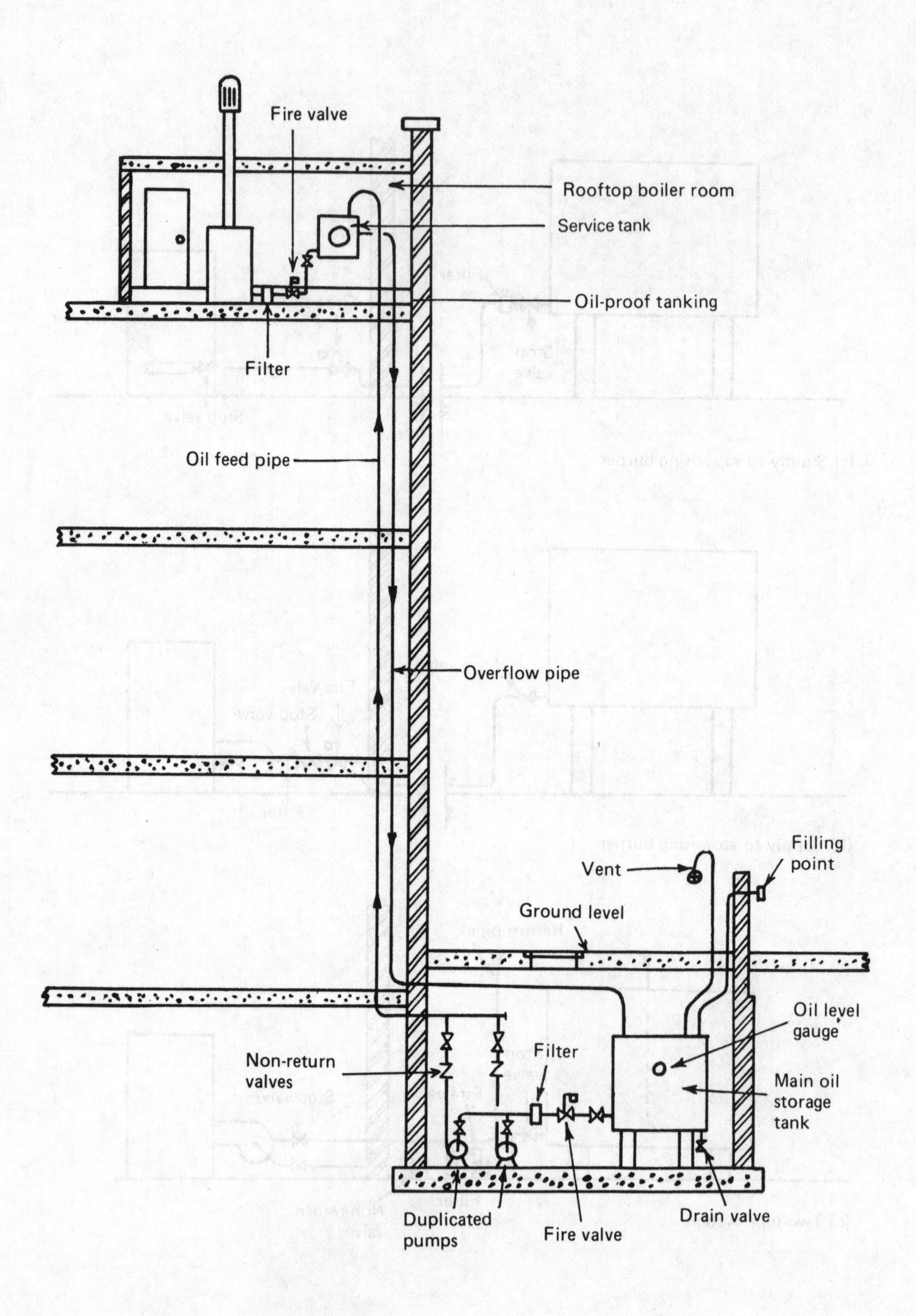

Fig. 8.11 Oil supply to rooftop boiler room.

sion wire incorporating a fusible link (see Fig. 8.12). The fusible link is usually sited above the boiler burner and in the event of a fire the link breaks, the tension wire released, and the valve closes under the action of the spring or weight. This type of fire valve is usually suitable for use only in separate boiler houses, since the tension wire cannot usually be installed in a house.

2. A pressure-operated type, comprising a bellows-type valve connected by a capillary tube to a heat-sensitive bulb or phial, which is positioned where it will detect excessive temperature. In the event of a fire, the bulb is overheated, causing an increase in vapour pressure in the system, which expands the bellows inside the valve causing it to close (see Fig. 8.13). This type of fire valve is usually used for domestic oil-fired installations.

Combustion Efficiency Testing

The efficiency of combustion can be assessed by taking measurements of

(a) carbon dioxide (CO_2) content of the flue gases,
(b) temperature of the flue gases,
(c) solids in the flue gases,
(d) flue draught.

Although combustion efficiency testing is applied here to oil-fired installations, the same instruments and methods can also be applied to gas installations. A table for the combustion efficiency for gas, however, would have to be used and this would be supplied by the manufacturer of the testing equipment.

Carbon Dioxide Indicator (Fig. 8.14). This consists of a glass flask containing a liquid reagent and a hand-operated syringe-type pump. To measure the CO_2 content of the flue gases, the plunger valve is first depressed with the indicator upright and the fluid level at zero on the scale. The probe is then inserted into a sampling hole in the flue and the plunger is depressed six times to warm the line and to eliminate air. A sample of the flue gas is then taken by depressing the plunger fully and squeezing the bulb of the pump at least 20 times, ensuring that the plunger remains fully depressed. The connecting plunger is then removed and the indicator inverted twice to

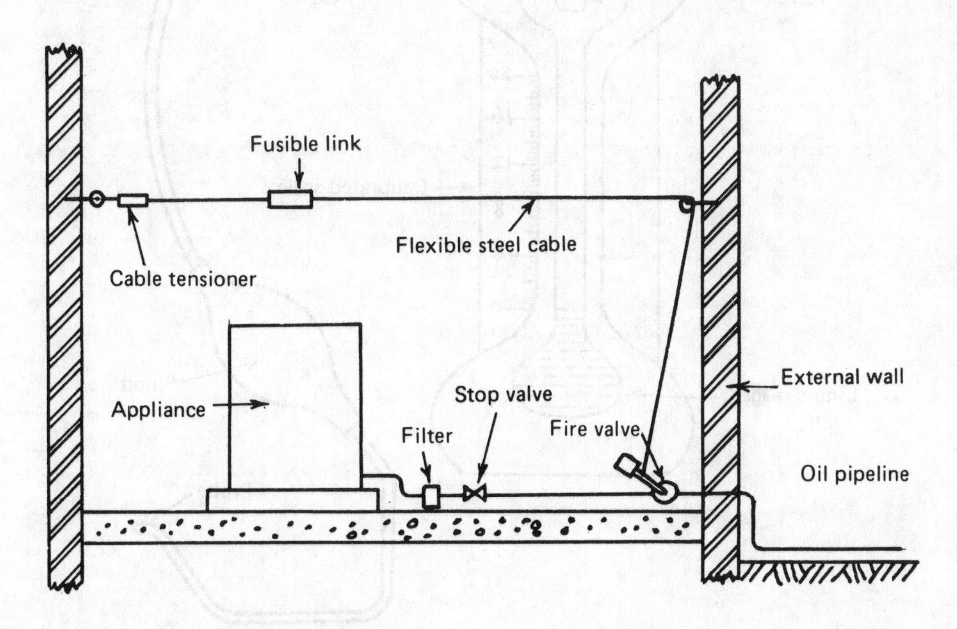

Fig. 8.12 Fusible-link-operated fire valve.

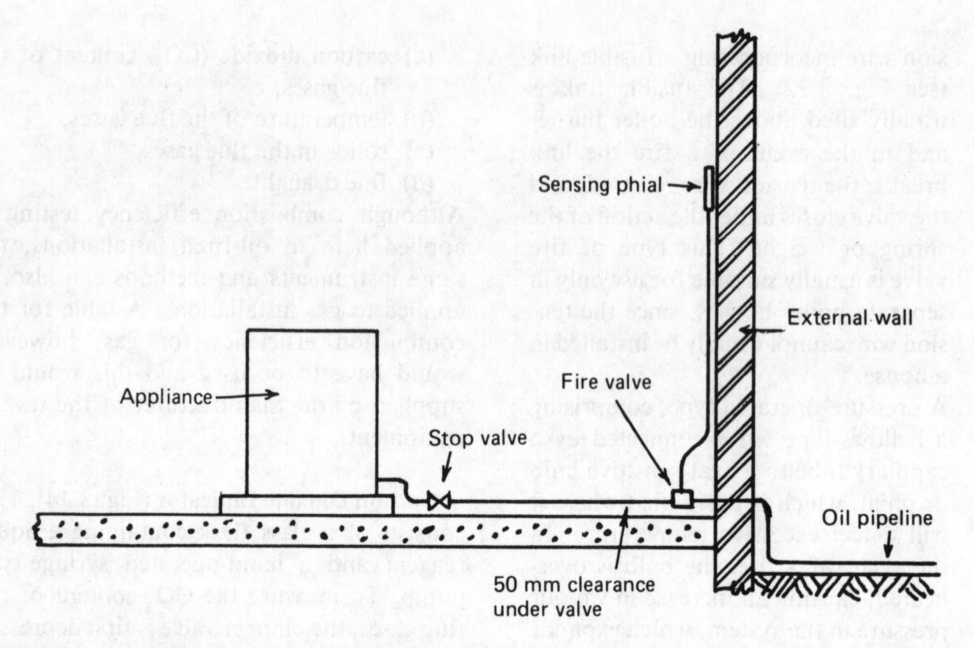

Fig. 8.13 Vapour-pressure-operated fire valve.

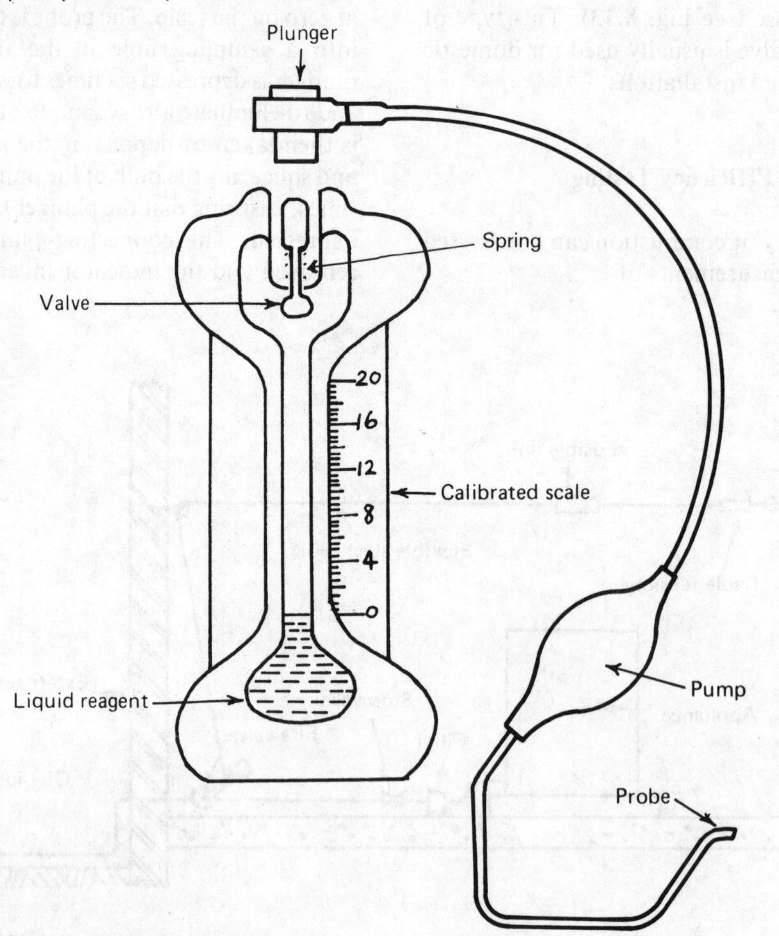

Fig. 8.14 Carbon dioxide indicator.

ensure that the sample of gas is fully absorbed by the liquid.

Absorption of CO_2 creates a partial vacuum inside the instrument, so the fluid is forced up the centre of the tube by an amount equal to the CO_2 absorbed. The percentage of CO_2 in the flue gas is then read off the scale against the level of the liquid in the tube. The instrument should continue to be inverted and the readings taken again until a consistent value is obtained.

Temperature of Flue Gases is taken by a thermostat shown in Fig. 8.15.

Solids in flue gases. The instrument used (Fig. 8.16) consists of a metal suction pump and a flexible tube terminating with a probe. A transverse slot in the pump body takes a special filter paper. To take a reading, the pump is first warmed to prevent condensation of the flue gases as they are sampled. A clean filter paper is inserted in the slot and the probe inserted in a test hole in the flue. A sample of flue gas is then drawn through the filter paper with ten full strokes of the pump. The filter paper is then withdrawn and the smoke deposit compared with the smoke scale in Fig. 8.17.

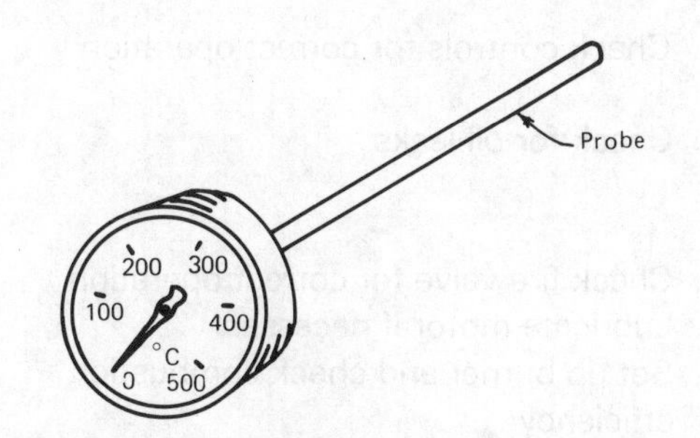

Fig. 8.15 Flue gas thermometer.

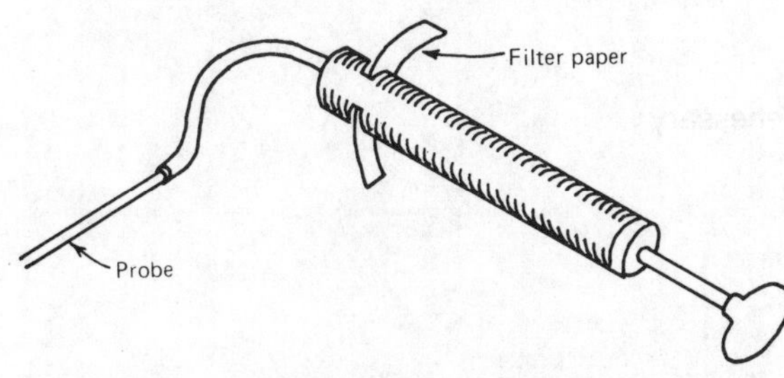

Fig. 8.16 Smoke tester.

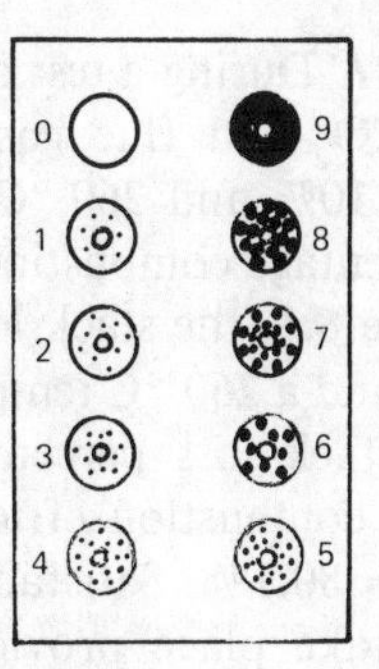

Fig. 8.17 Smoke scale.

Flue Draught. The instrument used is similar in appearance to the flue gas thermometer except that the probe is a hollow tube and the dial reads Pa instead of °C. The draught has an important bearing on burner efficiency and its value in Pa should equal that specified by the manufacturer of the appliance. Note: the appliance should be run for about 15 min before testing.

Combustion Performance

The values of CO_2 content and stack temperature can now be used by means of Table 8.1 to find the percentage combustion efficiency. From the table, the stack loss percentage figure must be deducted from 100 to give the combustion efficiency.

Table 8.1 Combustion Efficiency

% CO_2	Stack loss (%) at					
	93°C	144°C	204°C	260°C	316°C	371°C
14	9.9	12.1	14.3	16.7	18.7	21.0
13	10.1	12.4	14.8	17.4	19.6	22.0
12	10.3	12.8	15.3	18.0	20.3	22.9
11	10.5	13.2	16.0	18.9	21.4	24.2
10	10.8	13.8	16.8	19.9	22.6	25.6
9	11.3	14.4	17.7	21.1	24.1	27.4
8	11.7	15.3	18.9	22.6	26.0	29.6
7	12.3	16.3	20.4	24.6	29.4	32.6
6	13.2	17.7	22.4	27.2	31.6	36.5
5	14.3	19.8	25.3	31.0	36.2	42.0
4	16.1	22.8	29.6	36.5	43.0	50.0

Example 8.1. During a test on an oil-fired boiler, the CO_2 and flue gas temperature reading were 10% and 260 °C respectively. Find the percentage combustion efficiency.

From Table 8.1, the stack loss for a 10% CO_2 reading and a 260 °C temperature reading gives a stack loss percentage of 19.9; therefore the combustion efficiency will be 19.9 − 100 = 80.1%. Satisfactory combustion should take place providing that the smoke number is three or less when burning a domestic fuel oil having a relative density of 0.7 and five or less when burning a domestic fuel oil having a relative density of 0.8.

There are two grades of oil for domestic burners:

1. Class C2 or kerosene,
2. Class D or gas oil (similar to fuel used in diesel engines).

Regular maintenance checks of oil-fired

Table 8.2 Equipment Service Maintenance Schedule

The boiler-burner unit

Vaporising	*Atomising*
1. Inspect the chimney flue-ways, advise the customer if they require sweeping	1. Inspect the chimney flue-ways, advise the customer if they require sweeping
2. Clean the boiler flue-ways	2. Clean the boiler flue-ways
3. Clean burner or pot and renew wicks where necessary	3. Check the combustion chamber refractories and make good where necessary
4. Clean and check electrical ignition and associated controls if necessary	4. Remove nozzle assembly, clean nozzle and electrodes, replace and reset
5. Clean oil feed pipes to burner and check flexible connections	5. Inspect filters, clean or replace where necessary
6. Check filter and check operation of controls	6. Check controls for correct operation
7. If fan fitted, check motor operation: clean impeller blades and air inlet guard where necessary	7. Check for oil leaks
8. Check for oil leaks	8. Check fire valve for correct operation
9. Check fire valve for correct operation	9. Lubricate motor if necessary
10. Set up burner and check combustion efficiency	10. Set up burner and check combustion efficiency
11. Leave the installation in a clean condition	11. Leave the installation in a clean condition

Oil storage

1. Check tank for sludge and drain off where necessary
2. Check tank gauge for correct indication

Heating system

1. Check pump for correct operation
2. Check thermostats for correct operation
3. On warm-air systems, check operation of air distribution fan and air filters

appliances should be made on all mechanical and electrical equipment in accordance with the manufacturers instructions. During the two routine maintenance visits, service will be carried out as shown in Table 8.2.

Chimneys for Oil-Fired Boilers

If the products of combustion of the oil inside the chimney are allowed to cool to the region of the acid dew point temperature, some of the acid will remain inside the chimney and may attack the material from which it is constructed.

Unburnt particles of carbon are also present in the products of combustion and may collect on any rough surfaces or points of change of velocity. When the boiler is fired, the sudden flow of the products of combustion may cause these carbon deposits to be discharged to the atmosphere. To prevent the occurrence of these problems, the following points must be observed.

1. The chimney should be well insulated and lined with acid-resisting material.
2. Bends should be avoided, and if they have to be included they should have a large radius.
3. The draught stabiliser should not permit excessive cool air to enter the chimney.
4. Where two or more boilers are installed, each boiler should have its own chimney or, if a common chimney is unavoidable, the flues connecting the boilers to the chimney must be provided with dampers, so that an idle flue may be closed from the others.

Note: a balanced flue may also be used for oil-fired boilers.

Draught Stabilisers (Fig. 8.18)

A well designed and constructed chimney will often produce excessive draught and to maintain the correct draught for optimum conditions of firing a draught stabiliser is usually fitted at the base of the chimney.

The stabiliser consists of a hinged metal frame having a hinged flap. The hinged flap is set to maintain the required draught conditions and will automatically open to permit cool air to enter the chimney when the draught is too high. The hinged frame acts as an explosion door to relieve excessive pressure inside the chimney; should this occur, the hinged frame is forced open, carrying with it the hinged flap. After this has occurred, the hinged frame and flap automatically close.

Calorific Values of Fuels

The calorific value of a fuel may be defined as the heat energy produced per unit mass or per unit volume of the fuel when it is fully burned. In practice it is necessary to allow for heat losses when using calorific values.

Table 8.3 gives the approximate calorific values of some common fuels. If other factors are known, calorific values may be used to calculate the amount of fuel used for a heating system and the cost.

Example 8.2. Calculate the mass of fuel required per week for a house having heat losses of 20 kW. An oil-fired boiler having an efficiency of 80% and is to be fired for 10 h each day. Calorific value of oil = 45 MJ/kg.

$$\text{kW} = \frac{\text{kJ}}{\text{s}}$$

$$\text{kJ} = \text{kW} \times \text{s}$$

$$\text{kJ} = 20 \times 7 \times 10 \times 60 \times 60$$

$$\text{kJ} = \frac{20 \times 7 \times 10 \times 60 \times 60 \times 100}{80}$$

Table 8.3 Calorific Values of Fuels

Fuel	Calorific value (MJ/kg)
Coal	30
Wood	18.5
Anthracite	32
Coke	25
Fuel oil (domestic)	45
Natural gas	36 MJ/m^3
Electricity	3.6 MJ/kW

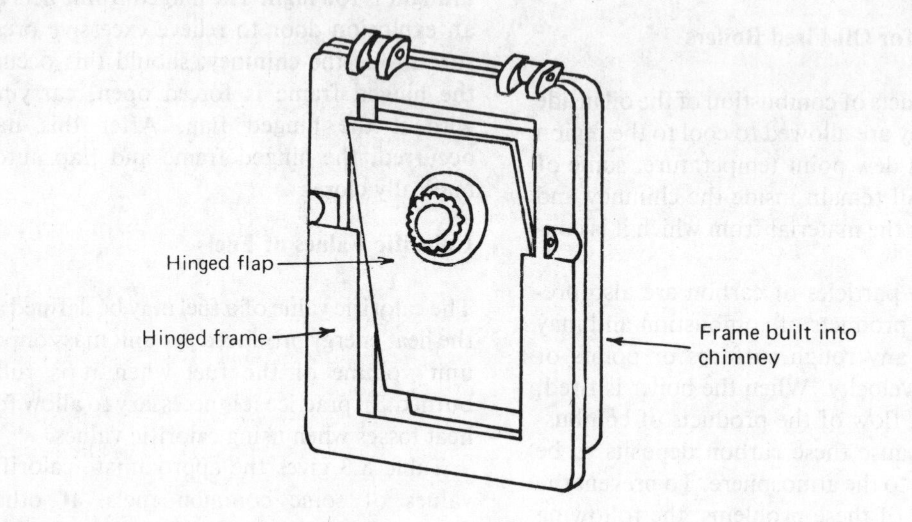

(a) View of draught stabiliser

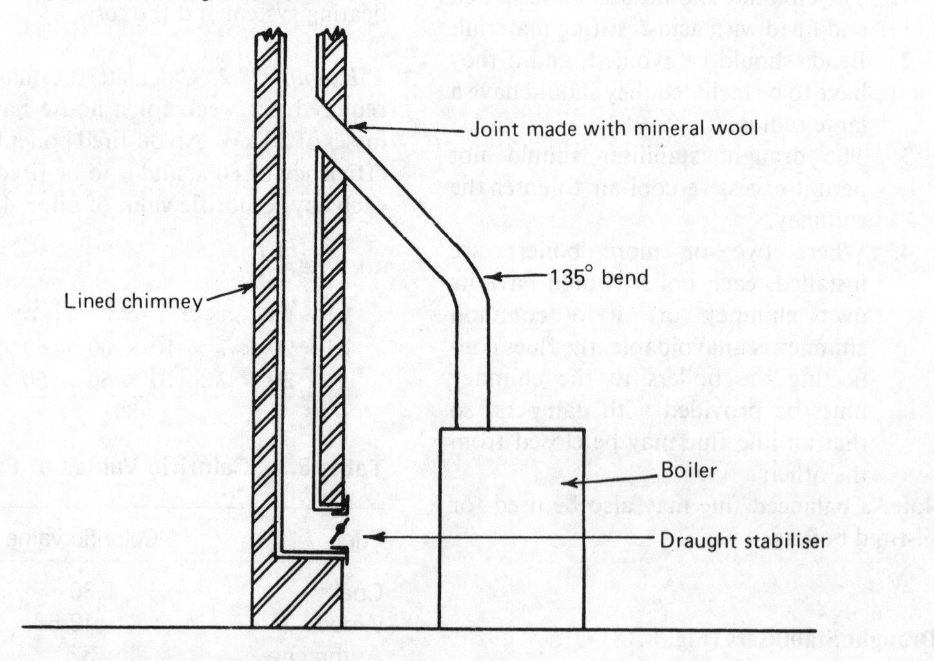

(b) Installation of draught stabiliser

Fig. 8.18 Draught stabiliser.

$$= 6\ 300\ 000$$
$$MJ = 6300$$
$$\text{mass} = \frac{MJ}{\text{calorific value (MJ/kg)}}$$
$$= \frac{6300}{45}$$
$$= 140 \text{ kg}$$

Note: once the mass of fuel required is found, the cost may be easily found by multiplying the cost per kg by the mass required in kg. The cost of fuel increases rapidly, but assuming that the cost of 1 kg of oil is 5p, the running cost per week for the house heating system would be

$$= 5 \times 140$$
$$= 700\text{p}$$
$$= £7.00$$

Exercises

1. Explain why it is essential to vaporise or atomise oil before it can be burned.
2. Sketch and explain the operation of the following types of oil burner;
   (a) wallflame,
   (b) dynaflame.
3. Sketch and explain the operation of natural and forced-draught vaporising burners and state why it is necessary to provide a heat leak with these types of burner.
4. State the purpose of a constant level controller. Sketch a section through a double-float constant-level controller and explain its operation.
5. Sketch an elevation of a pressure-jet atomising burner and explain its operation.
6. State the usual capacities of oil tanks for small installations and the type of material required for their construction.
7. Sketch an isometric view of an oil-storage tank showing all the necessary controls and pipework.
8. State the usual positions for the siting of an oil storage tank.
9. Sketch the method of supplying oil to a rooftop boiler room.
10. Sketch a section through an oil tank room showing the method of installing the oil tank and the necessary fire precautions.
11. State the purpose of a fire valve. Sketch and explain the operation of the following types of fire valve:
    (a) mechanical,
    (b) liquid expansion.
12. Sketch and describe the method of carrying out a combustion efficiency test on an oil-fired boiler.
13. Sketch a chimney for an oil-fired boiler and state the types of material that may be used for its construction.
14. Sketch and explain the operation of a draught stabiliser.
15. Define the term 'calorific value of fuel' and state the calorific values of coal, anthracite, fuel oil, natural gas and electricity.

Chapter 9
Centralised Hot Water Supply, Pipe Sizing

Definitions

BOILER HEATING SURFACES:

(a) DIRECT: in which the surfaces of the boiler receive heat directly from the heat source,

(b) INDIRECT: in which the surfaces of the boiler receive heat indirectly from the fire, usually through internal flue ways.

BOILER-RATED OUTPUT: the power of the boiler rated in kilowatts.

CALORIFIER: a vessel in which water is heated by means of an internal heat exchanger.

COLD FEED PIPE: a pipe connecting the cistern to the hot water system.

COMBINED BOILER AND STORAGE VESSEL: a boiler unit on which is superimposed a storage vessel.

DEAD LEG: a single pipe for drawing off hot water from a calorifier, cylinder or tank.

FEED AND EXPANSION CISTERN: an open-top vessel for maintaining the water level in the system and for accommodating increases in the volume of water due to expansion.

HEAD TANK: an auxiliary hot water storage vessel with open vent, used to ensure an adequate supply of hot water to certain draw-off points.

HOT WATER CYLINDER: a closed cylindrical vessel in which hot water is stored.

HOT WATER TANK: a closed rectangular vessel in which hot water is stored.

INDIRECT CYLINDER: a term usually used to define a small domestic storage calorifier.

ISOLATING VALVE: a valve used to isolate a circuit.

NOMINAL CAPACITY: the total volume of a cistern, tank or cylinder, calculated from the external dimensions of these vessels.

OPEN VENT: an open pipe from any high point in a hot-water system or from any closed vessel containing hot water.

PRIMARY CIRCUIT: the flow and return pipes through which hot water circulates between the boiler (or other heating appliance) and the storage vessel.

SECONDARY CIRCUIT: the flow and return pipes through which hot water circulates between the storage vessel and the hot-water draw off points.

Centralised Systems

In a centralised system, water is heated and stored centrally and distributed throughout the building by means of pipework. The heating of the water is usually achieved by means of a boiler which may be fired by gas, oil or solid fuel. The hot water is stored inside a cylinder or calorifier and these should be sited as close as possible to the boiler, so that the heat losses fromthe primary circuit is reduced to a minimum. A combined boiler and storage vessel would reduce these heat losses to the absolute minimum.

The hot water storage vessel should be well lagged and, in large systems for hospitals, schools, factories, etc., the primary and secondary circuits should be lagged.

A centralised system has the following advantages.

1. The hot water storage vessel contains sufficient water to meet a large anticipated peak demand load. This is important for large buildings where bulk supplies are required.
2. It may be possible to use cheaper fuel, i.e. lower-grade oil or solid fuel.
3. The risk of a fire inside the building is greatly reduced and confined to the boiler room.
4. One boiler plant reduces maintenance.

There are two distinct types of system: direct and indirect.

Direct Systems. In these, the water in the boiler circulates direct to the hot water storage vessel without passing through a heat exchanger. The system is cheap to install and the direct circulation provides a quick heat-recovery period. If soft water is used, the boiler must be rustproofed or rusty water will be drawn off through the taps. In temporary hard water districts there is a risk of scaling of the boiler and primary circuit, and the water temperature must therefore never exceed 60 °C, otherwise excessive scaling will occur.

Indirect Systems. In these, the water in the boiler circulates through a heat exchanger fitted inside the hot water storage vessel and the water is heated indirectly. If steam is available this may be circulated through the heat exchanger instead of hot water.

Since the water in the boiler does not mix with the water in the storage vessel, in soft water districts there is no risk of rusty water being drawn off through the taps. Because the water in the boiler, primary circuit and heat exchanger is not drawn off through the taps, there is no risk of scaling when temporary hard water is used.

After initial heating of the water and precipitation of the carbonates, there should be no further occurrences of this precipitation. Indirect systems have a further advantage when central heating is combined with the hot water supply, because there is no risk of rusty water from the radiators being drawn off through the taps. The water temperature may also be higher than that used for the direct system, which is necessary for the central heating system.

Domestic Installations. In domestic buildings the water is heated by a floor or wall-mounted boiler, or by a back boiler behind the living room grate. Fig. 9.1 shows a direct system to supply bath, basin, sink, shower and towel airer, and Fig. 9.2 shows an indirect system for supplying the same fittings.

Primatic Cylinder (Fig. 9.3)

A saving in the cost of an indirect system can be achieved by the use of a single-feed cylinder. The cylinder is provided with a patented heat exchanger, which is designed to provide two airlocks between the primary and secondary waters. These airlocks prevent mixing of the two waters and the cylinder thus acts in the same way as a conventional indirect cylinder with a heating coil. The cylinder dispenses with an expansion and feed cistern, primary feed pipe and primary vent pipe. Fig. 9.4 shows the installation of the primatic cylinder: although it is an indirect system, the system is installed similar to the direct system.

Sealed Primary Circuits

If a conventional indirect cylinder is required, an alternative method of dispensing with the expansion and feed cistern involves the use of a sealed primary circuit. To provide an expansion space for the heated water, an expansion vessel must be fitted to the circuit, usually on the return pipe where the water temperature is lower. Fig. 9.5 shows a cistern-fed indirect system, which is usually permitted in the UK.

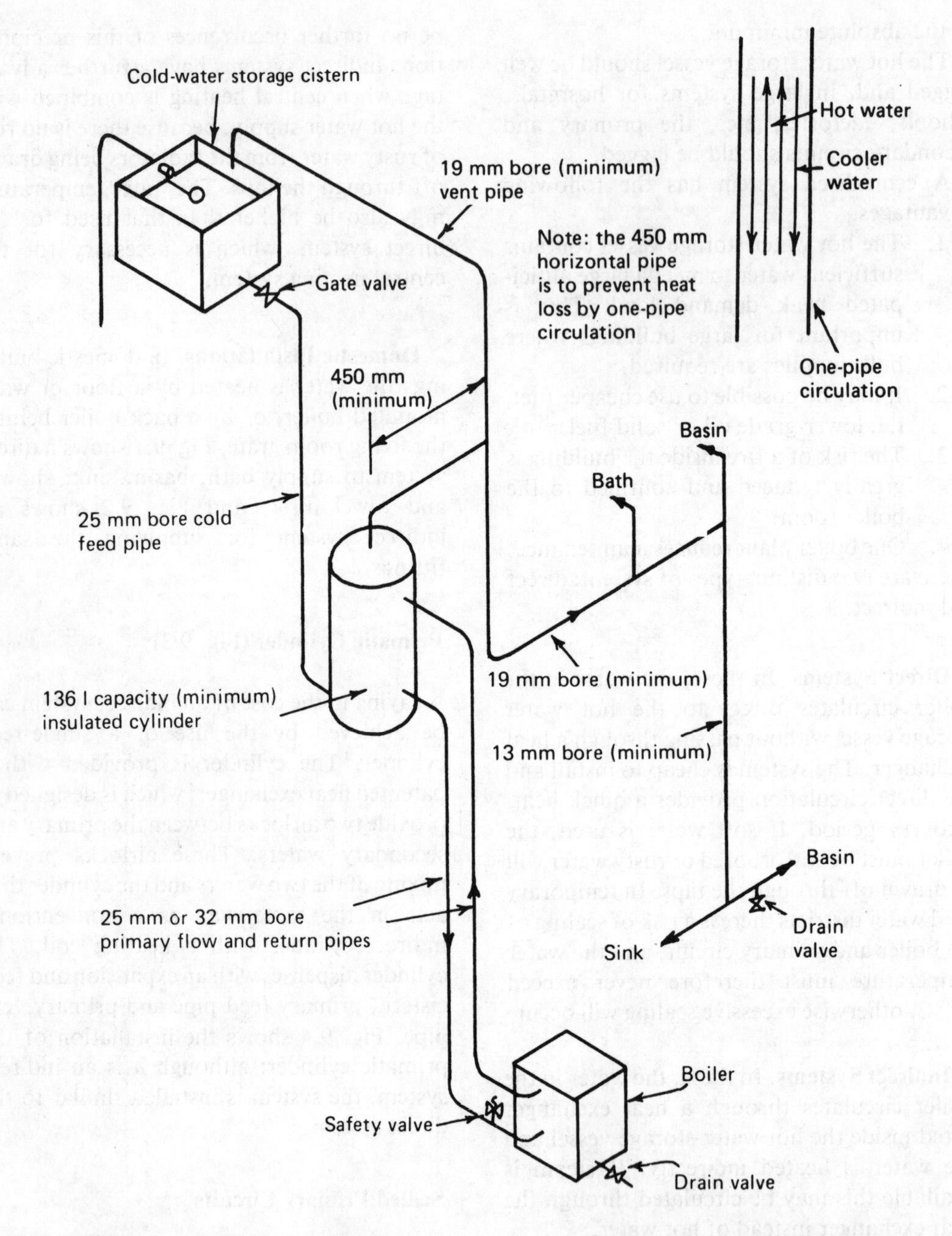

Fig. 9.1 Direct hot-water supply system.

The Building Regulations 1985 permit the use of unvented hot water storage systems. The secondary circuit is fed directly from the water main, which dispenses with the cold water storage cistern, secondary cold feed and vent pipe.

Because the hot water storage vessel is subjected to pressure from the water main, the metal will have to be thicker than for a cistern-fed storage vessel. As an alternative to the use of thicker metal, a pressure-reducing valve may be installed on the cold water ser-

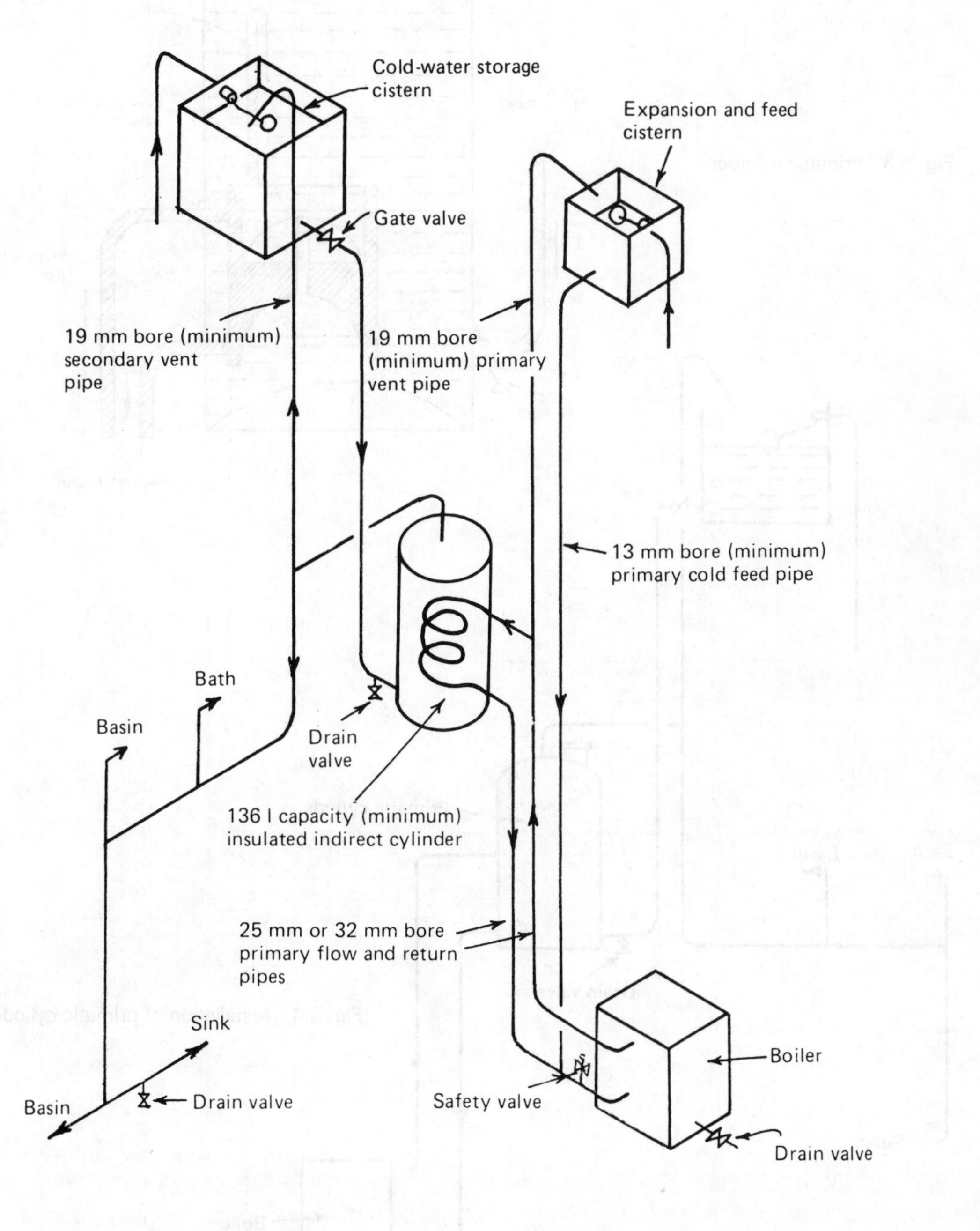

Fig. 9.2 Indirect hot-water supply system.

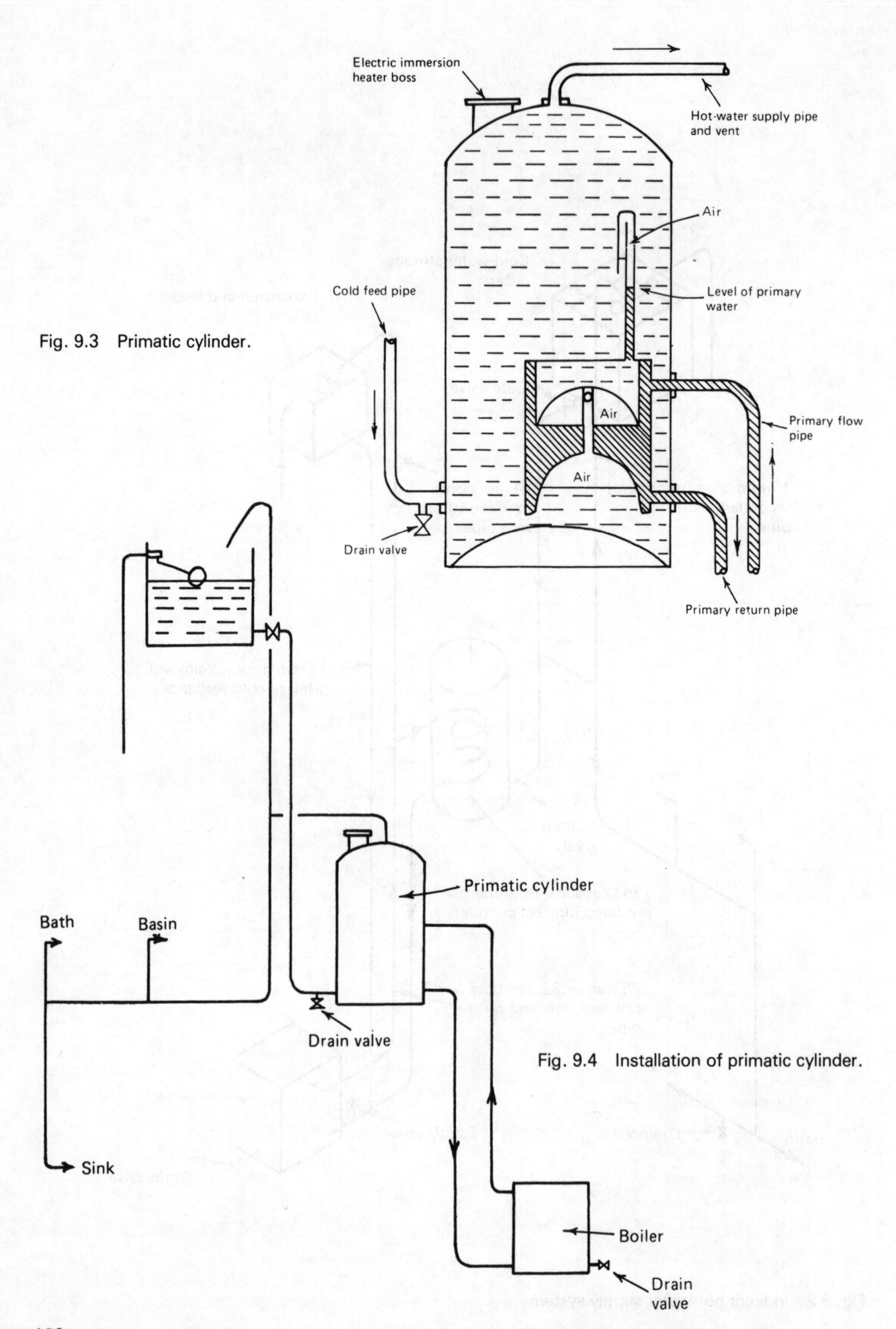

Fig. 9.3 Primatic cylinder.

Fig. 9.4 Installation of primatic cylinder.

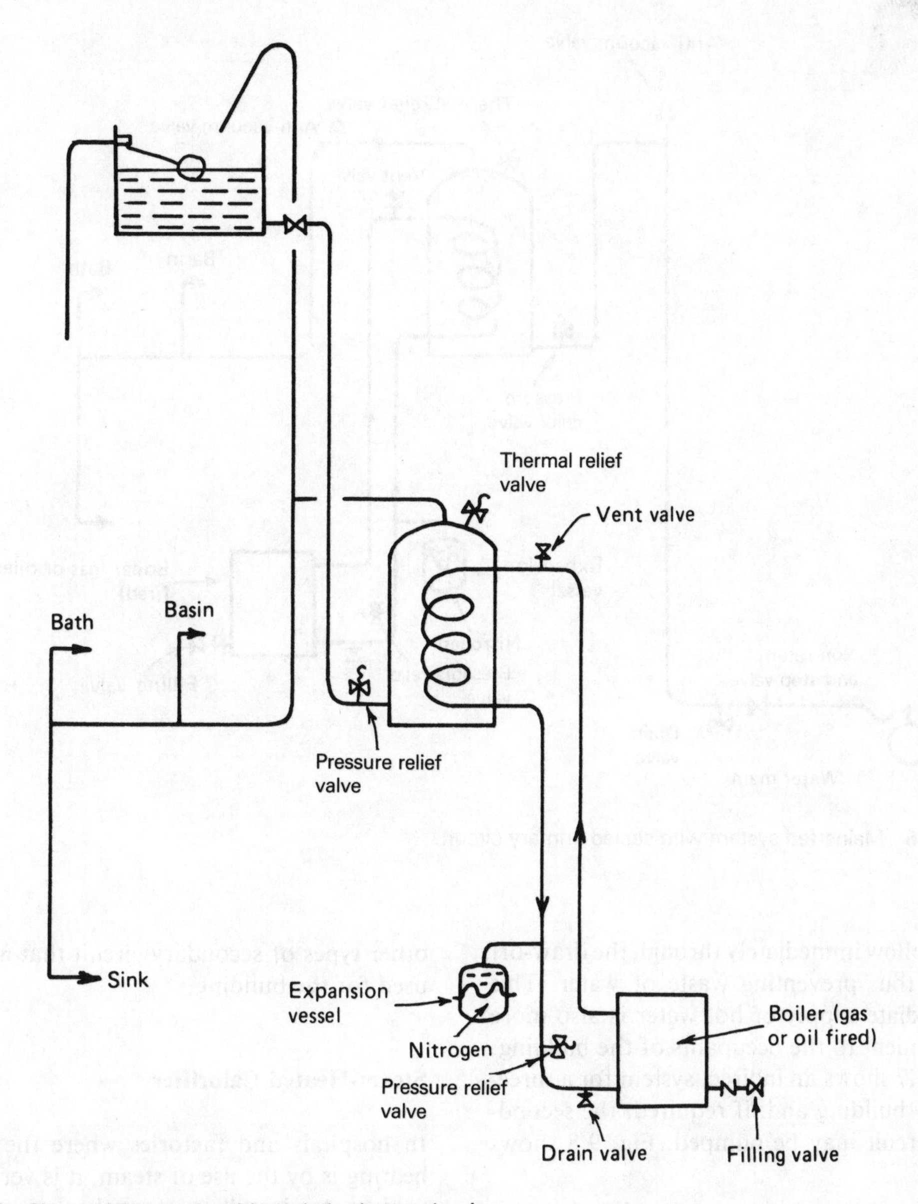

Fig. 9.5 Cistern-fed system with sealed primary circuit.

vice pipe supplying the vessel.

Fig. 9.6 shows a mains-fed system with a sealed primary circuit. The anti-vacuum valve is a precaution against siphonage of water from the cylinder and draw-off taps back into the water main.

Systems for Large Buildings

Because the local water authorities prevent the use of long 'dead legs', it is necessary in a large building to provide a secondary return back to the hot water calorifier. A dead leg can result in a great deal of water being wasted because each time water is drawn off through the taps cooler water in the pipe must first be run to waste. Table 9.1 gives the maximum lengths of hot water draw-off pipes not requiring insulation.

Since hot water is allowed to circulate when a secondary return is installed, the hot water

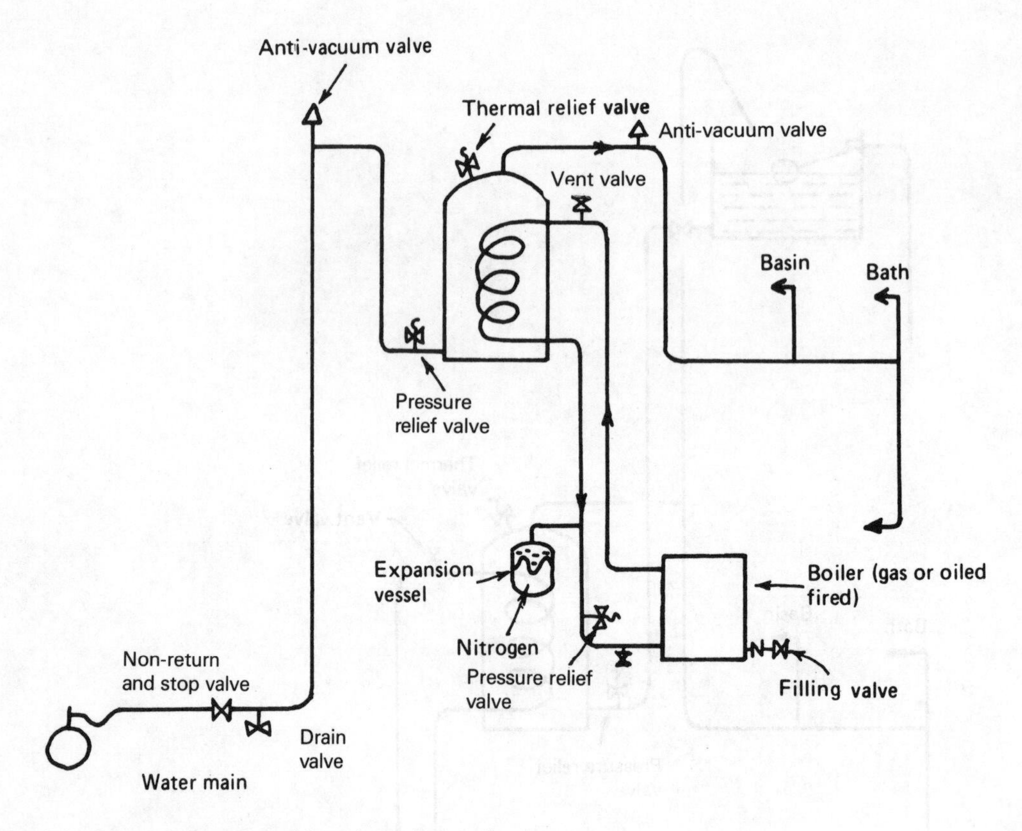

Fig. 9.6 Mains-fed system with sealed primary circuit.

will follow immediately through the draw-off taps, thus preventing waste of water. This immediate supply of hot water is also more convenient to the occupants of the building. Fig. 9.7 shows an indirect system for a three-storey building and, if required, the secondary circuit may be pumped. Fig. 9.8 shows

Table 9.1 Building Regulations 1985 for the Maximum Lengths of Hot Water Draw-off Pipes not Requiring Insulation

Outside diameter of pipe (mm)	Maximum length of uninsulated pipe (m)
Not more than 12	20
More than 12 but not more than 22	12
More than 22 but more than 28	8
More than 28	3

other types of secondary circuit that may be used for the building.

Steam-Heated Calorifier

In hospitals and factories where the space heating is by the use of steam, it is very convenient to install a steam-heated storage calorifier to provide hot water for supplying the ablution fittings. Fig. 9.9 shows the installation of a horizontal steam-heated storage calorifier.

Supplies to High-Rise Buildings (Fig. 9.10)

To prevent excessive water pressures on the lowest draw-off points, and consequent risk of water hammer and wear on valves, it is essential that the head of water acting on these points does not exceed 30 m. In build-

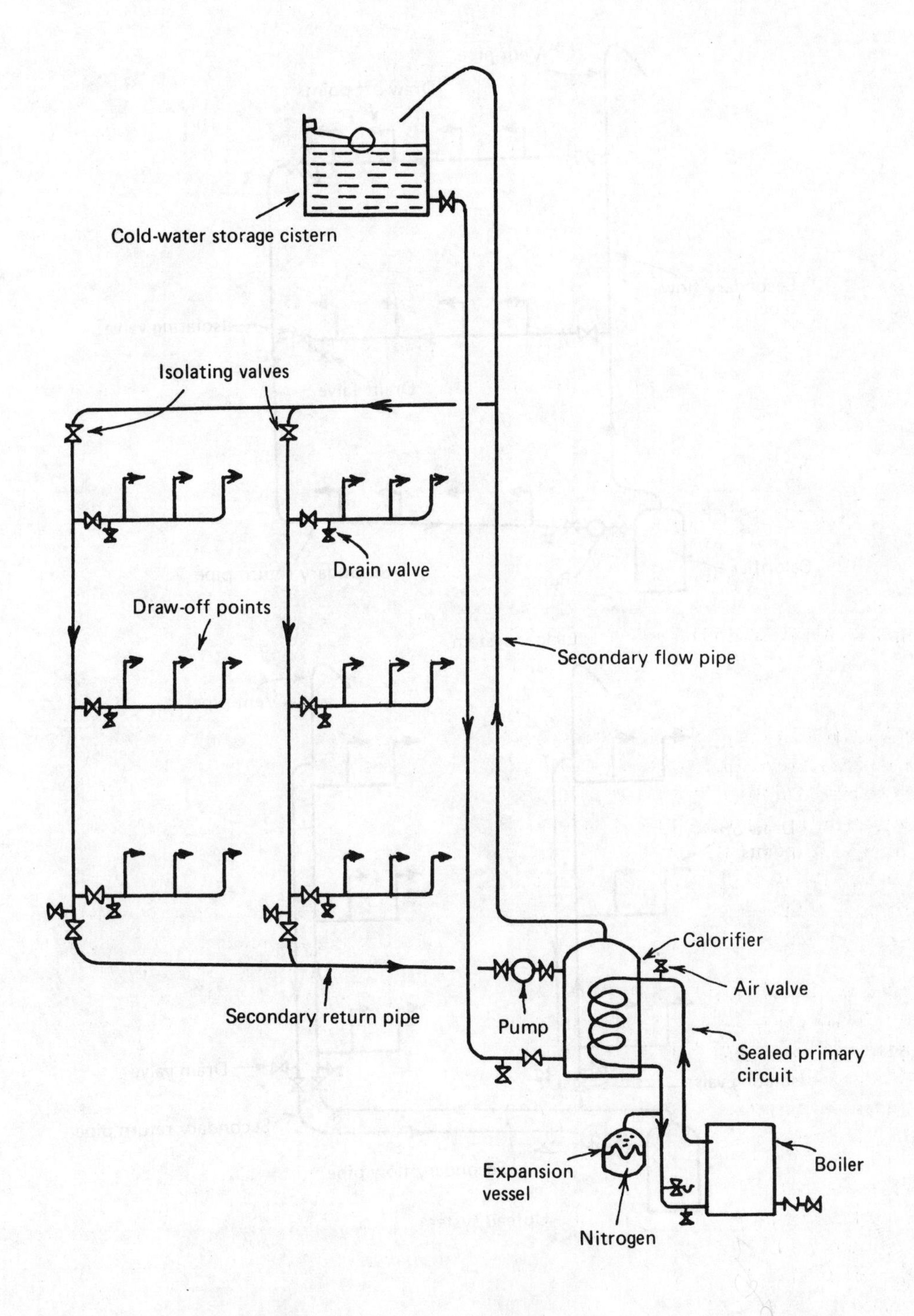

Fig. 9.7 Hot-water supply system for a three-storey building.

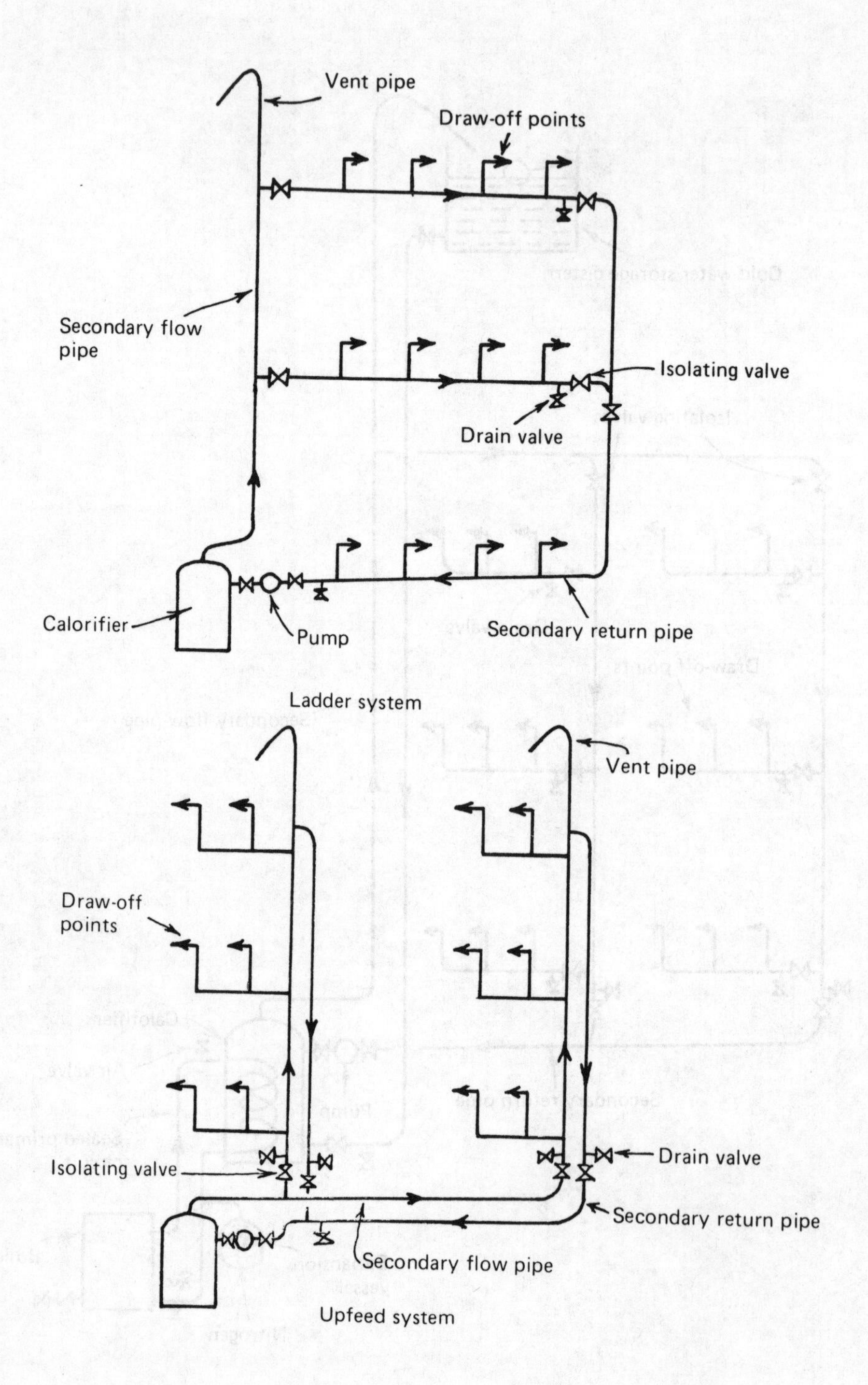

Fig. 9.8 Secondary circuits.

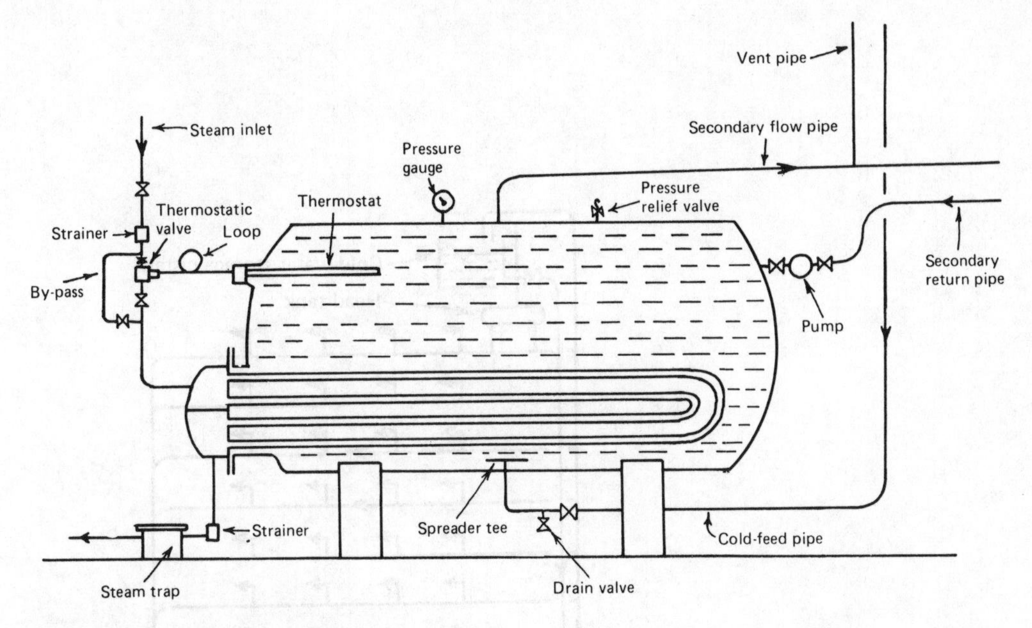

Fig. 9.9 Steam-heated storage calorifier.

ings above 30 m in height, therefore, two or more secondary circuits will be required. Each circuit must be supplied with water from a separate cold water storage cistern. The primary circuit may be sealed as shown, or fed from an open expansion and feed cistern.

Stand-By Plant (Fig. 9.11)

In buildings where it is essential to maintain a supply of hot water at all times, a stand-by plant will have to be installed.

The insertion of isolating valves as shown will allow either a boiler or a calorifier to be repaired or replaced whilst still maintaining a supply of hot water from the remaining plant. The three-way vent valve will ensure that the primary circuit is open to the atmosphere at all times, even when the stop valves are closed.

Solar Heating of Water (Fig. 9.12)

The heating of water by use of a flat-plate solar collector sited on the roof may save up to 40% of the heating cost of hot water supply. For a house, the collector should have an area of between 4 m^2 and 6 m^2 and be fixed at an angle of about 40 ° facing south.

Solar Collector. This consists of an aluminium alloy or copper frame with a sealed glass cover. A copper radiator or copper pipes, painted black, are fixed inside the frame so that there is an air space of about 20 mm between the glass and the radiator or pipes.

Aluminium foil should be fixed behind the radiator or pipes and 75–100 mm thickness of insulation fixed behind the foil. Fig. 9.13 shows a section through a solar collector.

Operation of the System.

1. The flat-plate collector is heated by solar radiation, which in turn heats a mixture of water and antifreeze inside the radiator or pipes to a temperature of up to 60 °C. On a sunny day it is possible to heat by solar energy all the water required for a house.
2. The pump is switched on when the temperature of thermostat A exceeds the temperature of thermostat B by between 2 and 3 °C.
3. The water and antifreeze in the solar collector is pumped through the heat

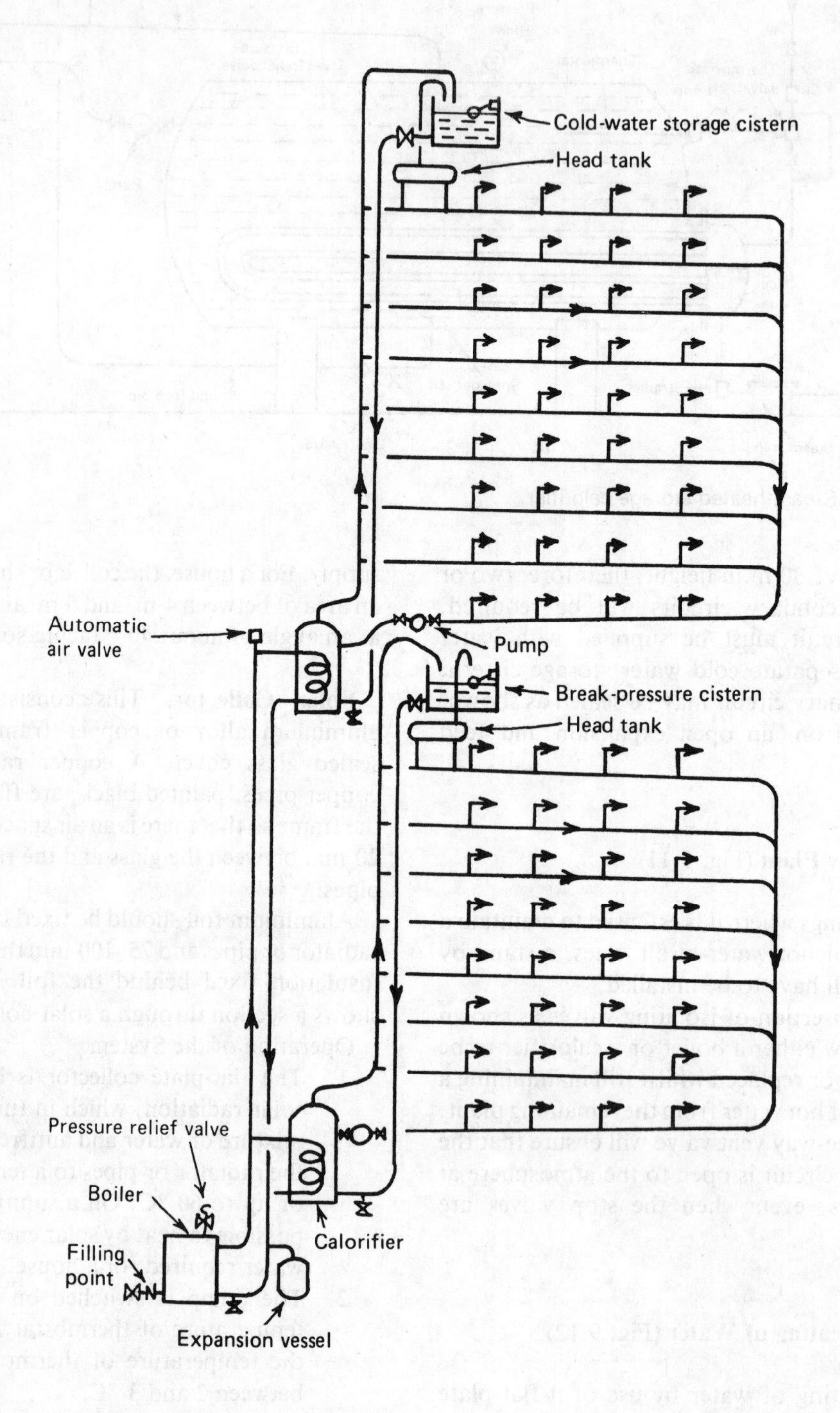

Fig. 9.10 High-rise building.

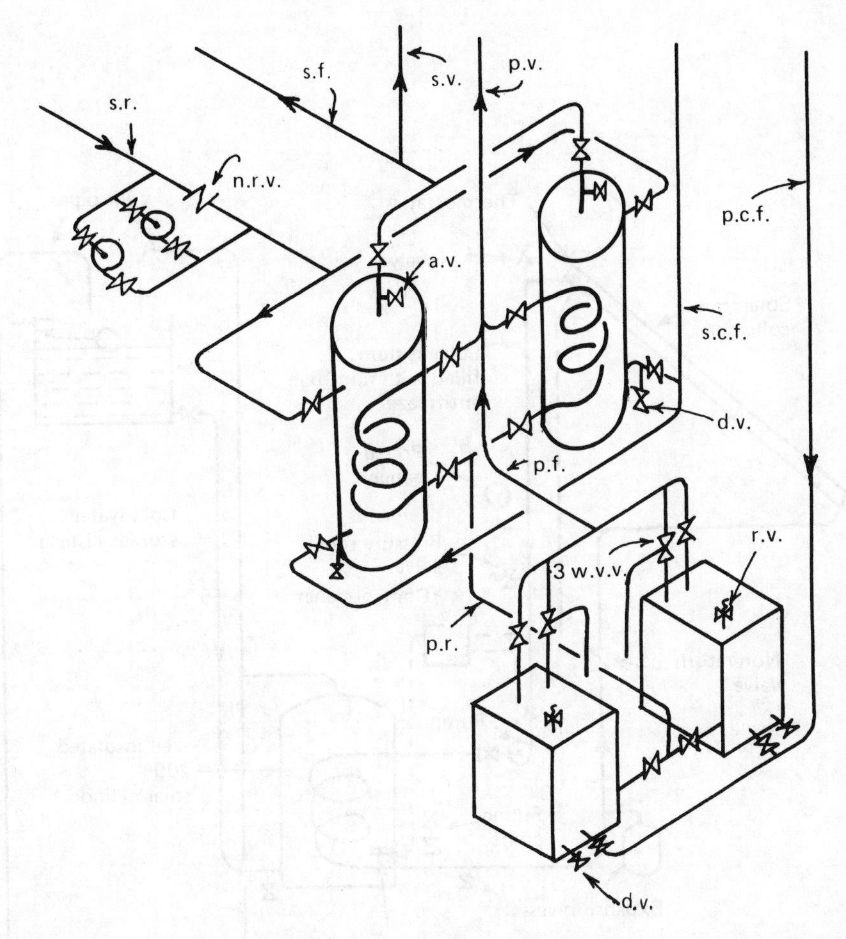

Fig. 9.11 Stand-by plant. s.r. = secondary return, n.r.v. = non-return valve, s.f. = secondary flow, s.v. = secondary vent, p.v. = primary vent, a.v. = air valve, p.f. primary flow, s.c.f. = secondary cold feed, d.v. = drain valve, 3w.v.v. = three-way vent valve, p.r. = primary return, r.v. = relief valve, p.c.f. = primary cold feed.

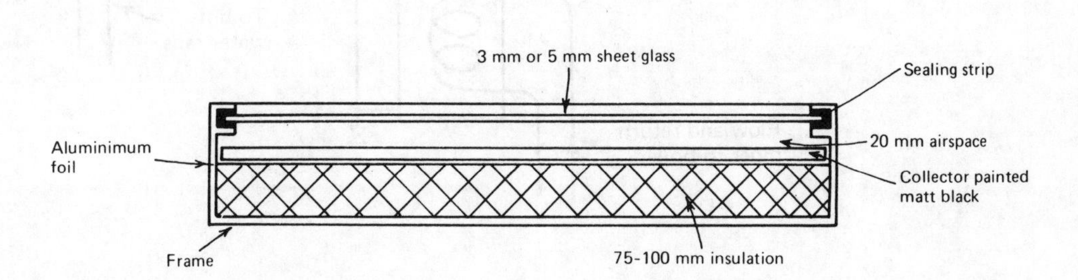

Fig. 9.13 Flat-plate solar collector.

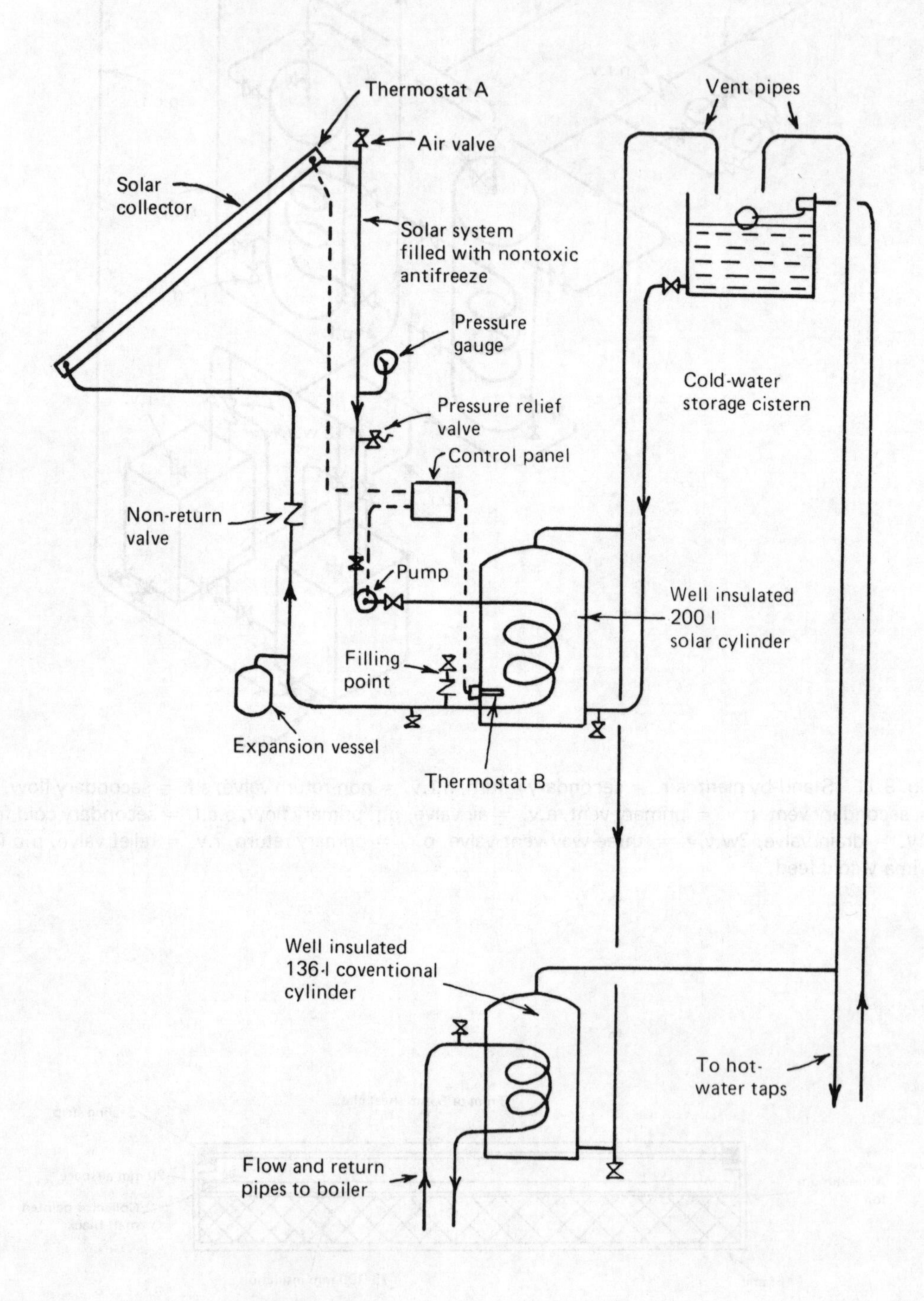

Fig. 9.12 Solar hot-water system.

exchanger inside the 200 l solar cylinder and the heat exchanger heats the water.
4. When hot water is drawn off through the hot water taps, the cold water from the cold water storage cistern forces hot water from the solar cylinder into the conventional cylinder and this reduces or eliminates the heat required to raise the temperature of the water in the conventional cylinder.

Note: to prevent loss of fluid in the solar circuit, a sealed system is used. The expansion vessel contains a rubber diaphragm with one side in contact with the fluid and the other side in contact with a cushion of nitrogen gas.

Hot Water Storage Requirement and Boiler Power

If the population of the building is known these may be calculated by use of Table 9.2.

Example 9.1. Calculate the hot water storage requirements and the boiler power for a factory having 250 employees, using Table 9.2.

$$\text{Storage requirement} = 250 \times 5 = 1250\ \text{l}$$

$$\text{Boiler power} = 250 \times 1.2 = 300\ \text{kW}$$

Combined Hot and Cold Water Unit (Fig. 9.14)

A factory-made self-contained unit for providing hot and cold water for sink, bath, basin and WC will save a great deal of work on site. The unit is based on a design developed by the Research and Development Group of the Ministry of Housing and Local Government, and is acceptable for the appropriate government improvement grant. It is readily handled upstairs and through doorways and may be in a cupboard, in a passage or convenient corner, or incorporated in the structure of new housing.

Table 9.2 Hot Water Storage and Boiler Power

Note: the temperature of the cold feed water to the hot water storage is assumed to be 10 °C. For other temperatures the boiler power will have to be increased or reduced as appropriate.

	Storage at 65 °C (l/person)	Boiler power to 65 °C (kW/person)
Colleges and schools		
boarding	25	0.7
day	5	0.1
Dwellings and flats		
low rental	25	0.5
medium rental	30	0.7
high rental	45	1.2
Factories	5	1.2
Hospitals		
general	30	1.5
infectious	45	1.5
infirmaries	25	0.6
infirmaries with laundry	30	0.9
maternity	30	2.1
mental	25	0.7
Nurses' homes	45	1.2
Hostels	35	0.9
Offices	5	0.1
Sports pavilions	35	0.3

Pipe Sizing for Hot and Cold Water

The internal diameter of a hot or cold water pipe required to discharge a given amount of water when subjected to a certain head or pressure of water may be found from tables or charts or by calculation.

A well known formula for the sizing of hot and cold water pipes was devised by Thomas Box and is expressed as follows:

$$q = \sqrt{\left(\frac{d^5 \times H}{25 \times L \times 10^5}\right)}$$

where q is the discharge through the pipe in l/s, d is the diameter of the pipes in mm, H is the head of water in m, and L is the effective length of pipe in m.

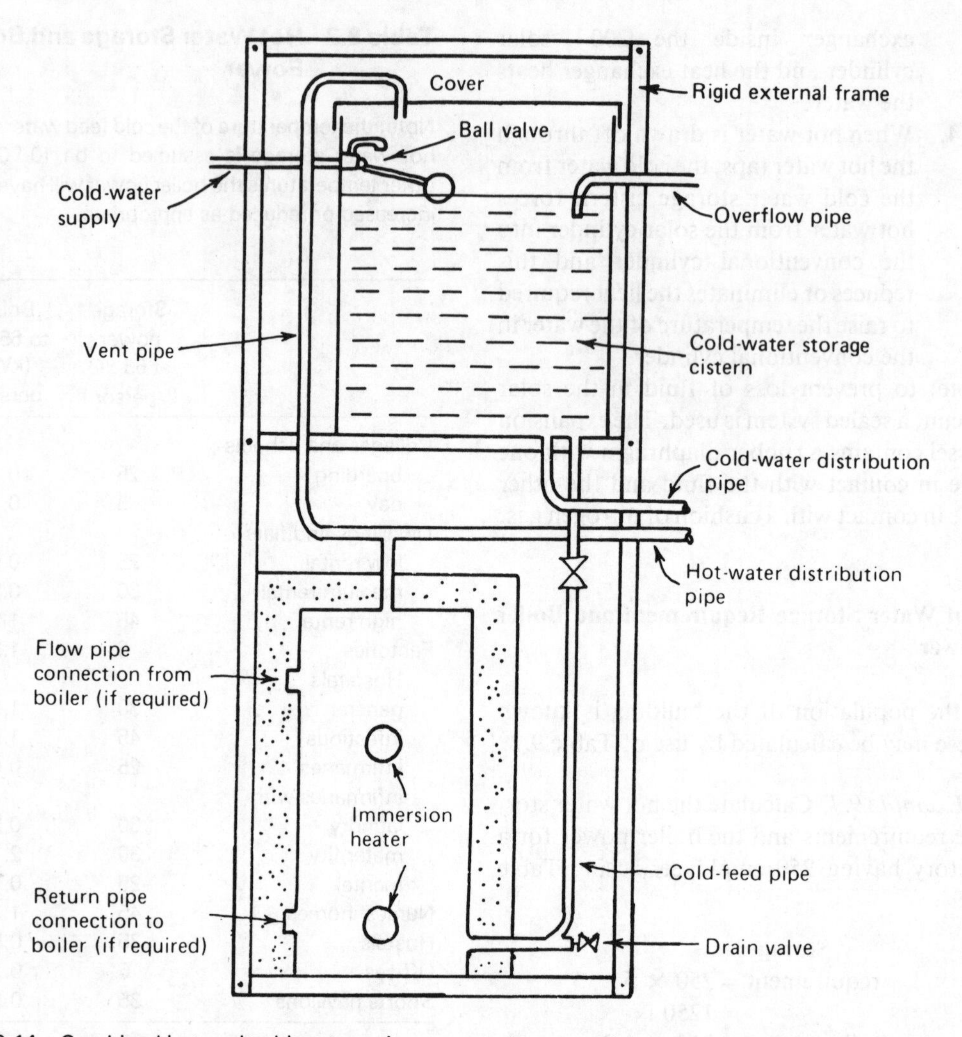

Fig. 9.14 Combined hot and cold water unit.

Example 9.2. Calculate the diameter of a pipe to discharge 0.2 l/s to a shower when there is a constant head of water of 1.5 m and an effective length of pipe of 5 m. Note: to find the diameter of the pipe, the Box's formula will have to be transposed to give an expression for diameter and this will be shown to help students unfamiliar with transposition. Most students possess a pocket calculator, but the ability to use logarithms is still required and logarithms have therefore been used in the calculation.

$$q = \sqrt{\frac{d^5 \times \mathrm{H}}{25 \times L \times 10^5}}$$

Squaring both sides,

$$q^2 = \frac{d^5 \times H}{25 \times L \times 10^5}$$

and cross-multiplying,

$$q^2 \times 25 \times L \times 10^5 = d^5 \times H$$

Dividing both sides by H,

$$d^5 = \frac{q^2 \times 25 \times L \times 10^5}{H}$$

and taking the fifth root of both sides,

$$d = \sqrt[5]{\frac{q^2 \times 25 \times L \times 10^5}{H}}$$

Therefore

$$d = \sqrt[5]{\frac{0.2^2 \times 25 \times 5 \times 10^5}{1.5}}$$

By logs:

Number	Log
0.04	$\bar{2}.6021$
25	1.3979
5	0.6990
10^5	5.0000
	5.6990
1.5	0.1761
(5th root) (÷ 5)	$\bar{5}.5129$
12.6	1.1025

A 13 mm bore pipe would be satisfactory.

Relative Discharging Power of Pipes

The relative discharging powers of pipes are as the square root of the fifth power of their diameters.

$$N = \sqrt{\left(\frac{D}{d}\right)^5}$$

where N is the number of branch pipes, D is the diameter of main pipe, and d is the diameter of branch pipes.

Example 9.3. Calculate the diameter of a water pipe to supply five washbasins each having a 13 mm diameter short branch pipe.

$$N = \sqrt{\left(\frac{D}{d}\right)^5}$$

squaring both sides,

$$N^2 = \left(\frac{D}{d}\right)^5$$

Taking the fifth root of both sides,

$$\sqrt[5]{N^2} = \frac{D}{d}$$

Multiplying both sides by d,

$$D = \sqrt[5]{N^2} \times d$$

Therefore

$$D = \sqrt[5]{5^2} \times 13$$

By logs,

Number	Log
25	1.3979
(5th root) (÷ 5)	1.3979
	0.2795
13	1.1139
24.74	1.3934

A 25 mm bore main pipe would be satisfactory.

Exercises

1. Define the following terms:
   (a) boiler heating surfaces,
   (b) boiler power,
   (c) calorifier,
   (d) dead leg,
   (e) feed and expansion cistern,
   (f) secondary circuit,
   (g) head tank.
2. State the principles of design of a centralised system of hot water supply.
3. Describe the direct and indirect systems of hot water supply and state the advantages of an indirect system over a direct system.
4. Sketch isometric diagrams of both direct and indirect systems of hot water supply for a house having bath, shower, basin, sink and towel rail.
5. State the advantages of a single-feed hot water cylinder and sketch a section through the cylinder and explain its operation.
6. Sketch a 'sealed' primary circuit and state its advantages over an 'open' primary circuit.
7. Sketch and annotate an isometric diagram of an indirect hot water system for a two-storey hotel having two bathrooms on each floor, each containing bath, basin and towel rail. The boiler and calorifier are to be sited on the ground floor adjacent to the bathrooms.
8. State the principles of design of a centralised hot water supply system for a high-rise building and sketch a system suitable for a 15-storey hotel having two

bathrooms on each floor, each containing bath, shower, basin and towel rail.

9. Sketch a section through a horizontal steam-heated storage calorifier showing all necessary valves and fittings.
10. Two boilers and two calorifiers are to be interconnected so that each appliance may be isolated for repair or renewal. Show by means of sketches the method of installing the boilers and calorifiers so that this can be achieved.
11. State the principles of design of a solar-heated hot water supply system. Sketch a solar system suitable for a house and explain its operation.
12. Calculate the hot water storage requirements and the boiler power for an office having 300 employees (Use Table 9.2.)
13. Calculate the diameter of a main pipe to supply thirty 19 mm diameter short branch pipes.
14. Calculate the head of water (in metres) required to discharge 5 l/s through a 50 mm bore pipe having an effective length of 45 m.
15. Calculate the number of 25 mm diameter branch pipes that may be supplied from a 76 mm diameter main pipe.
16. Calculate the diameter of a pipe to discharge 2 l/s when the effective length of pipe is 12 m and the head of water is 4 m.

Chapter 10
Localised Water Heating Systems

Principles

Instead of a centralised plant for the heating and storage of the hot water, localised systems, as the name implies, use electric or gas water heaters sited either directly over the fitting being supplied or as close as possible to the fitting. The lengths of the pipework and the heat losses from them are therefore reduced considerably and the cost of constructing a boiler room is eliminated. The heaters, however, take up a certain amount of wall or floor space within the building.

Electric Heaters

At the present time, the cost of heating water by electricity is higher than heating by other fuels and great care is required when designing a system in order to reduce the heat losses to the minimum. The following points must therefore be observed when installing a system using electric water heaters.

1. The lengths of the hot water supply pipes must be as short as possible, especially to the sink. The sink is normally used more than any other fitting and each time hot water is drawn off, hot water is left inside the pipe, which cools and loses heat.
2. The hot water storage vessel must be well insulated with a minimum of 50 mm thickness and preferably 76 mm of good-quality insulating material.
3. All circulation of electrically heated water must be avoided and therefore towel rails and radiators must not be installed on the system.
4. 'Single pipe' circulation in the hot water supply pipe or vent pipe must be avoided. In a system using a two-in-one pressure-type heater, this is achieved by carrying the hot water supply pipe from the top of the heater at least 450 mm horizontally before it connects to the vertical pipe as shown in Fig. 10.2.
5. The electric element must be controlled by a thermostat so that the water temperature does not exceed 60 °C in hard water areas and 70 °C in soft water areas.
6. Airing cupboards must not be heated by leaving part of the hot water storage vessel uninsulated. There is sufficient heat for airing linen from the exposed hot water pipes inside the airing cupboard.

Open-Outlet Single-Point Heaters (Fig. 10.1). All heaters of this type are designed to serve one fitting at a time. It is possible, however, to install the heater between two adjacent fittings so that the swivel open outlet can be turned to supply both fittings. Capacities vary from 1.5 to 68.1. The heater may be connected direct from the water main or a cold water cistern.

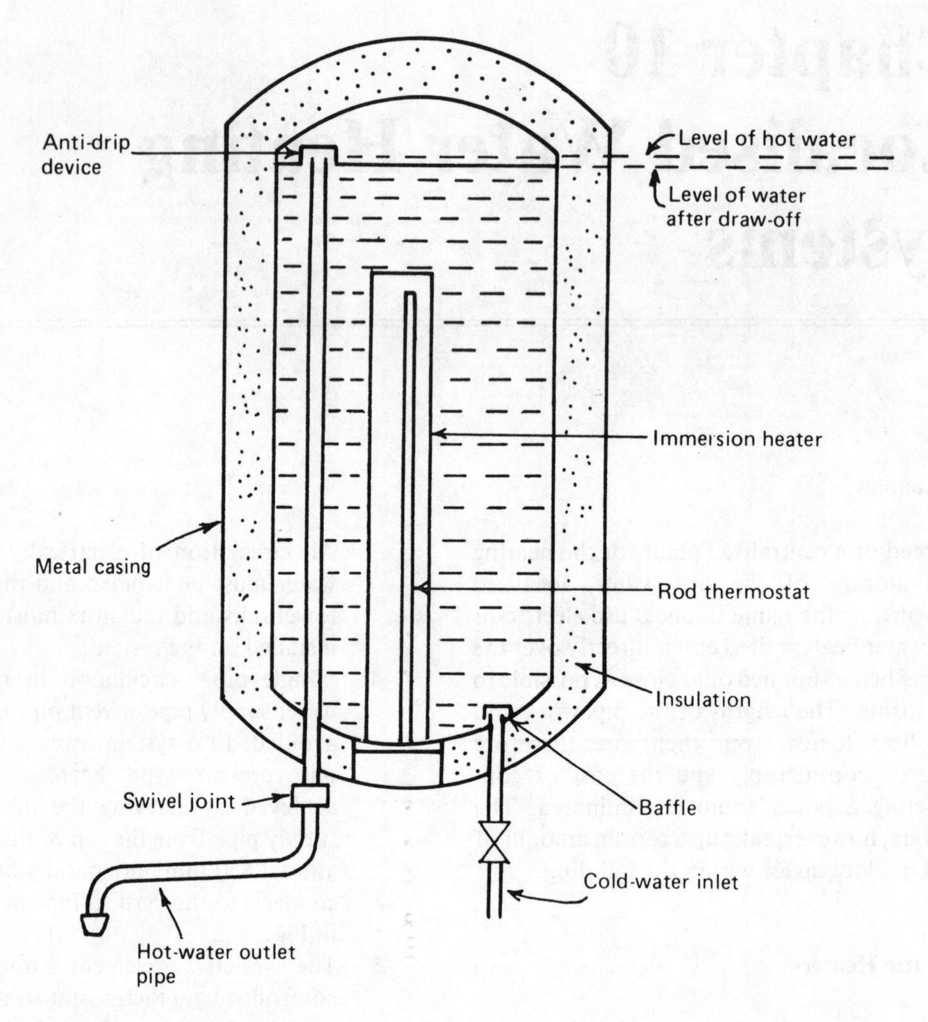

Fig. 10.1 Open-outlet electric water heater.

The baffle on the cold feed pipe prevents undue mixing of the cold water entering with the hot water inside the heater. The 'anti-drop' device prevents a drip of water through the outlet when the water is being heated and expands.

Pressure-Type Heaters (Fig. 10.2). These heaters are designed for use when hot water is required for two or more fittings. All models are totally enclosed and fed from a storage or feed cistern: they must never be supplied with cold water directly from the water main. The heater is normally fitted with two immersion heaters; one near the top, which is switched on for basin and sink use, and a larger heater at the bottom, which is switched on when larger quantities of water are required for bath and laundry use.

Fig. 10.3 shows the installation of a heater for a house. The heater is fitted near the sink to reduce the length of the hot water pipe. Fig. 10.4 shows the installation of a pressure heater for a three-storey building.

To prevent hot water from the heaters on the upper floors flowing down to the heaters below, the branch connections to the cold feed pipe must be taken above the heaters. The cold feed pipe must be vented at the top to prevent siphoning of hot water from the heaters on the upper floors down to the heaters on the lower floors. This could occur

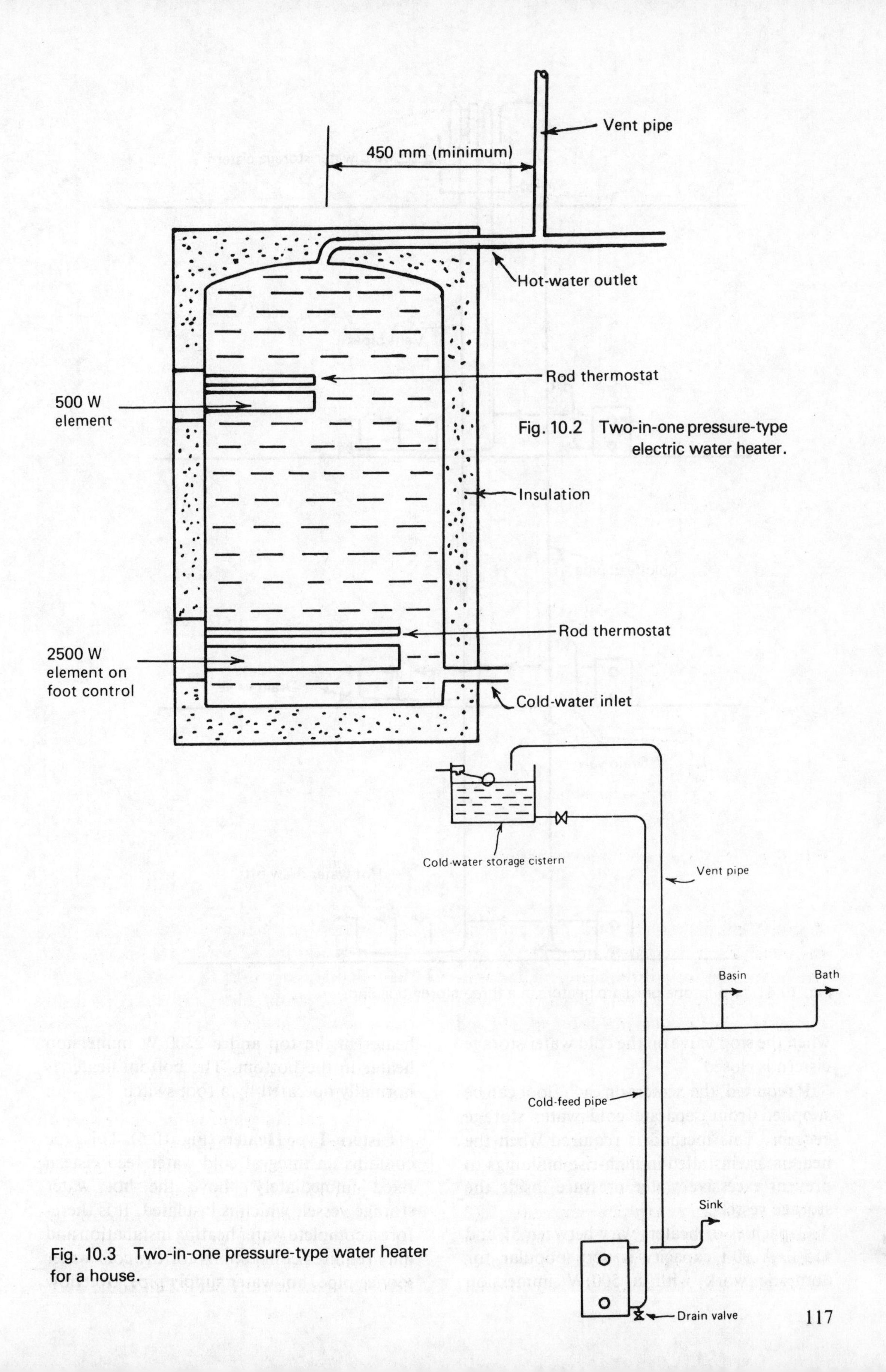

Fig. 10.2 Two-in-one pressure-type electric water heater.

Fig. 10.3 Two-in-one pressure-type water heater for a house.

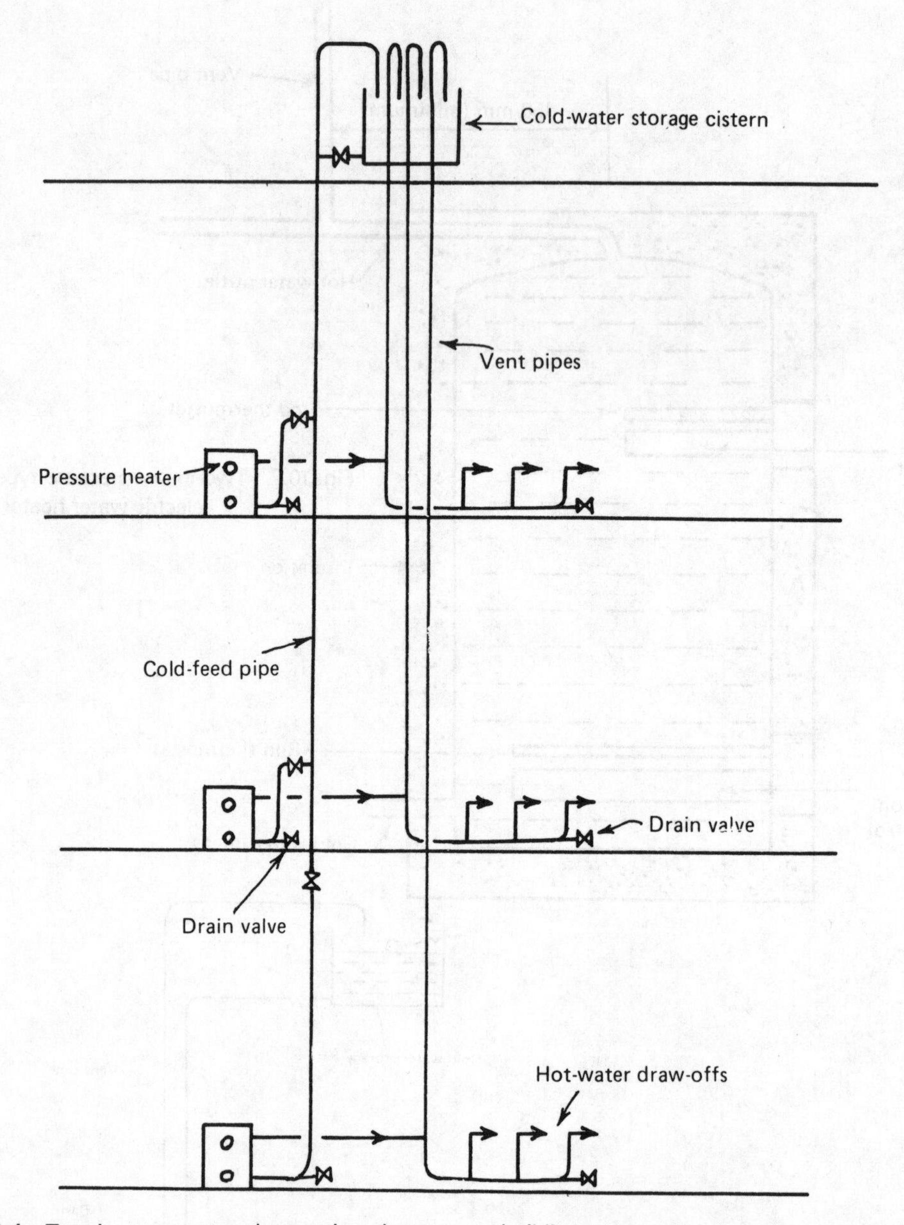

Fig. 10.4 Two-in-one pressure heaters in a three-storey building.

when the stop valve on the cold water storage cistern is closed.

If required, the heaters on each floor can be supplied from separate cold water storage cisterns. This method is required when the heaters are installed in high-rise buildings to prevent excessive water pressure inside the storage vessels.

Capacities of heaters vary between 55 and 455 l. A 90 l capacity is very popular for domestic work with a 500 W immersion heater in the top and a 2500 W immersion heater in the bottom. The bottom heater is normally operated by a foot switch.

Cistern-Type Heaters (Fig. 10.5). This type contains an integral cold water feed cistern fixed immediately above the hot water storage vessel, which is insulated. It is therefore a complete water heating installation and only requires connections for the cold water service pipe, hot water supply pipe, overflow

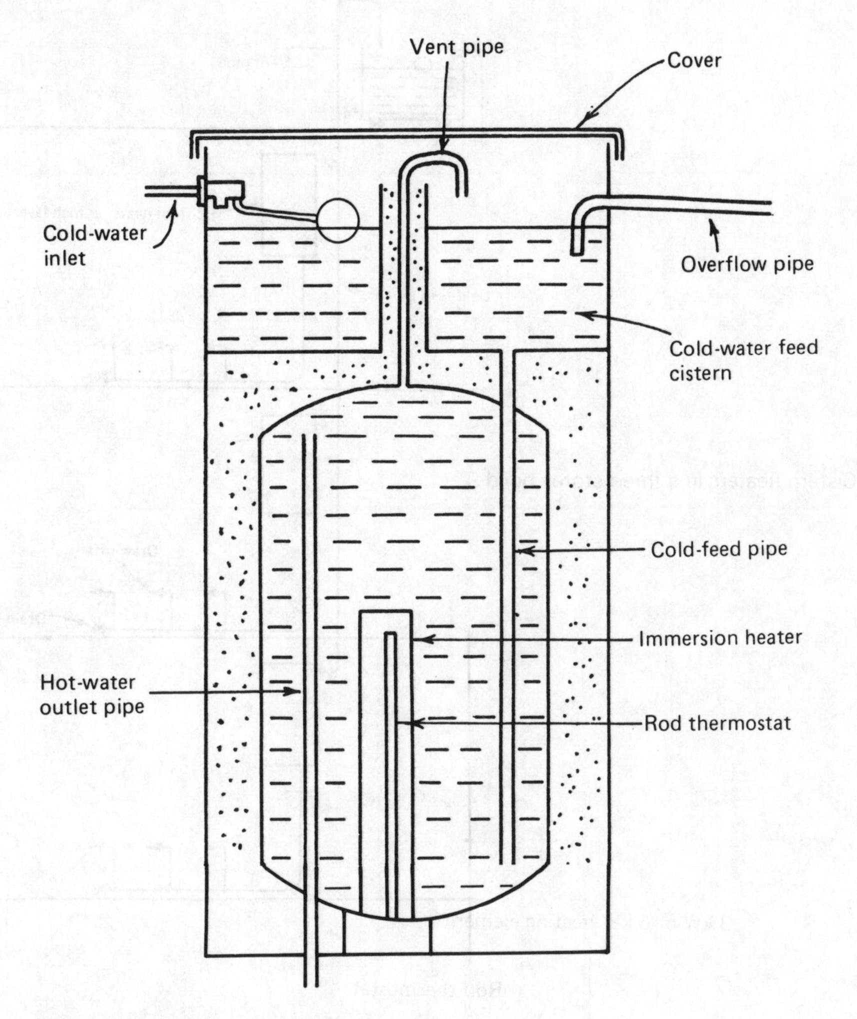

Fig. 10.5 Cistern-type electric water heater.

pipe and electrical service. Rectangular and circular models are manufactured for wall fixing, in capacities from 25 to 140 l of hot water storage. The base of the heater should always be higher than the highest hot water draw-off taps served.

Fig. 10.6 shows the installation of the cistern-type heaters for a three-storey building. The cold water supply to the heaters is shown connected to a cold water storage cistern. If the water mains pressure is sufficient, the heaters may be connected directly to the main.

Some units employ a variable-volume constant-temperature principle in which the cold water feed pipe is restricted to allow the electric immersion heater to heat the inflowing cold water quickly.

Instantaneous Heaters (Fig. 10.7). As the name implies, these heaters heat the cold water instantly as it flows through the heater, and storage of hot water is therefore not required. The heaters are much smaller than the other types of heater and are very popular for shower and washbasin use. The shower has an electric loading of 6 kW and will provide a continuous supply of warm water at showering temperature up to a maximum rate of approximately 3 l/min. The washbasin-type heater has an electrical loading of 3 kW and will provide a continuous supply of warm

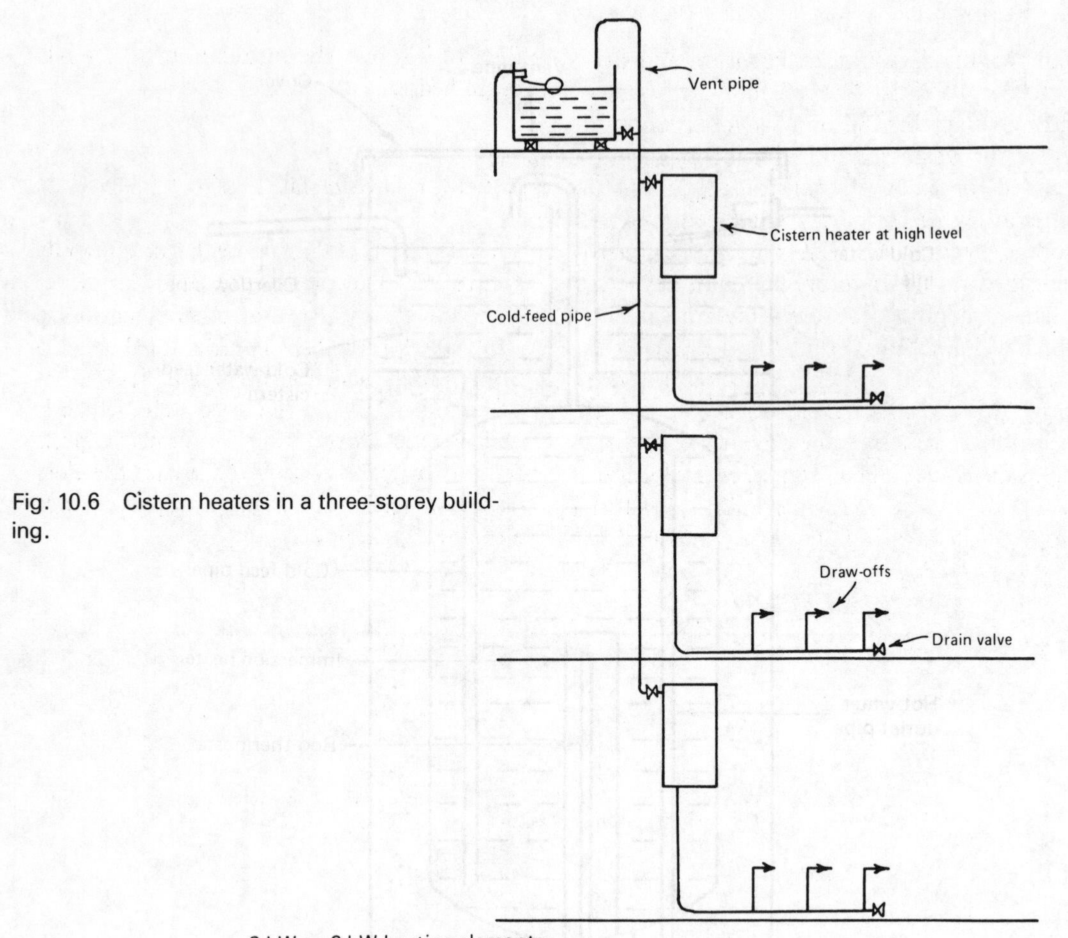

Fig. 10.6 Cistern heaters in a three-storey building.

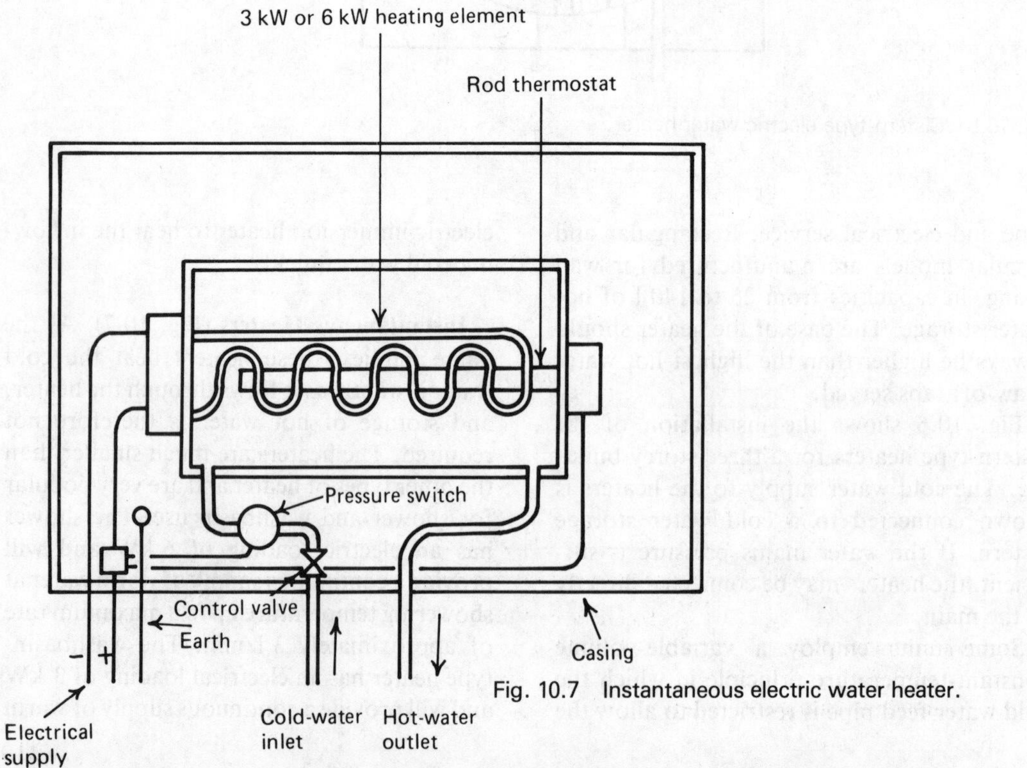

Fig. 10.7 Instantaneous electric water heater.

water for hand washing at the rate of approximately 1.4 l/min. The warm water passes through a spray head of a swivel arm and is ample for washing hands, but is not recommended for sinks, where greater amounts of water at higher temperatures are required. A storage type with open outlet previously described would therefore be required.

The heaters incorporate a pressure switch which will not allow the heating element to be switched on until water is flowing and vice versa. A preset thermal cut-out switch is also incorporated as a safeguard against overheating of the water. The heaters are usually connected directly to the water main. Fig. 10.8 shows the installation of a shower heater and Fig. 10.9 shows the installation of a washbasin heater.

Rod-Type Thermostat (Fig. 10.10)

All heating elements must be thermostatically controlled. The thermostat used for water heating consists of rod of invar steel (nickel steel) which expands very little, fixed inside a tube of brass which has a greater coefficient of expansion. When the brass tube, which is in contact with the water, expands, it pulls out the invar rod with it and breaks the electrical contact. The permanent magnet ensures a snap-action contact.

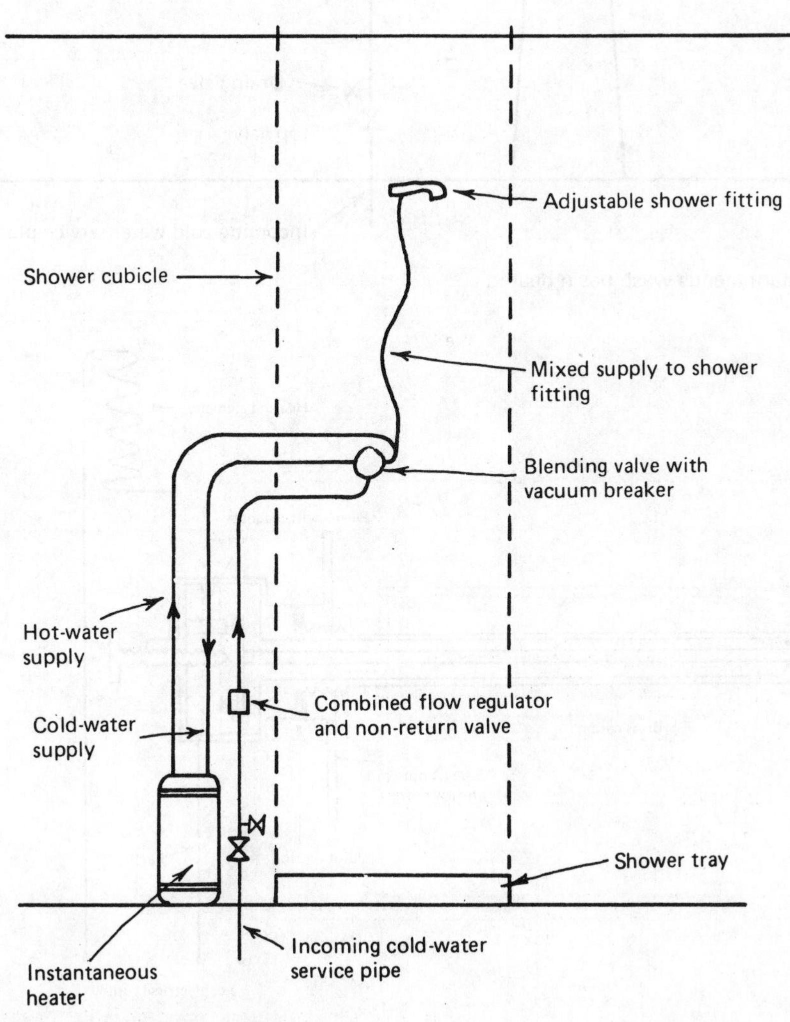

Fig. 10.8 Instantaneous shower heater.

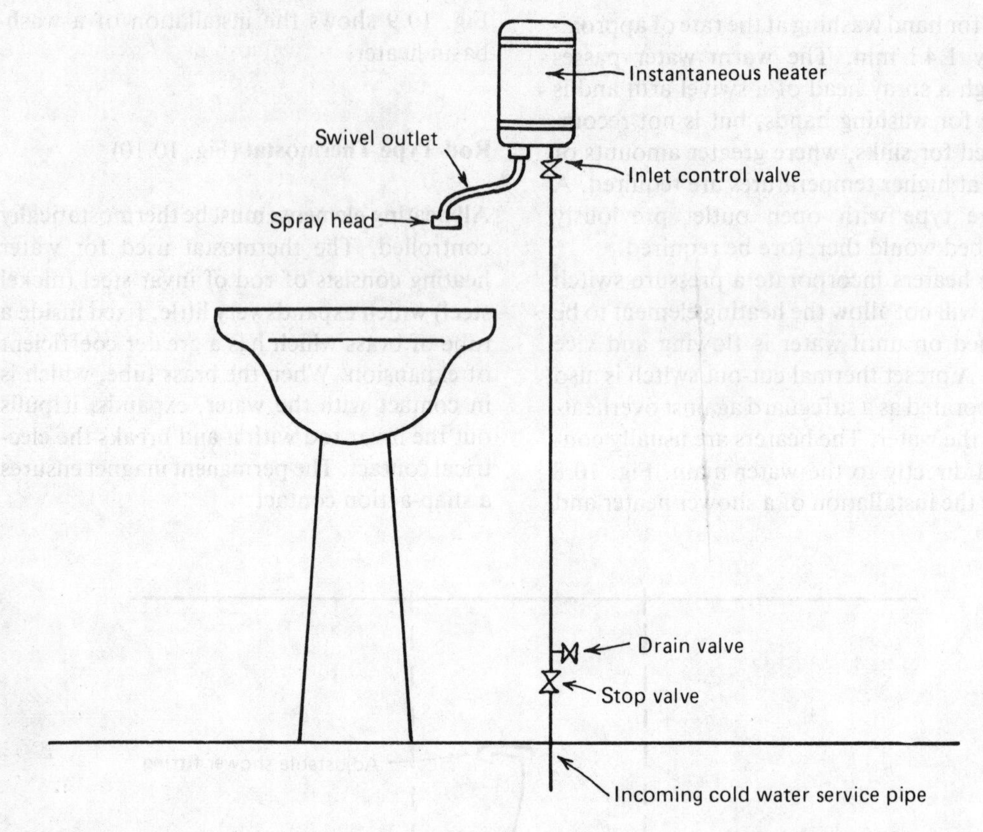

Fig. 10.9 Instantaneous wash basin heater.

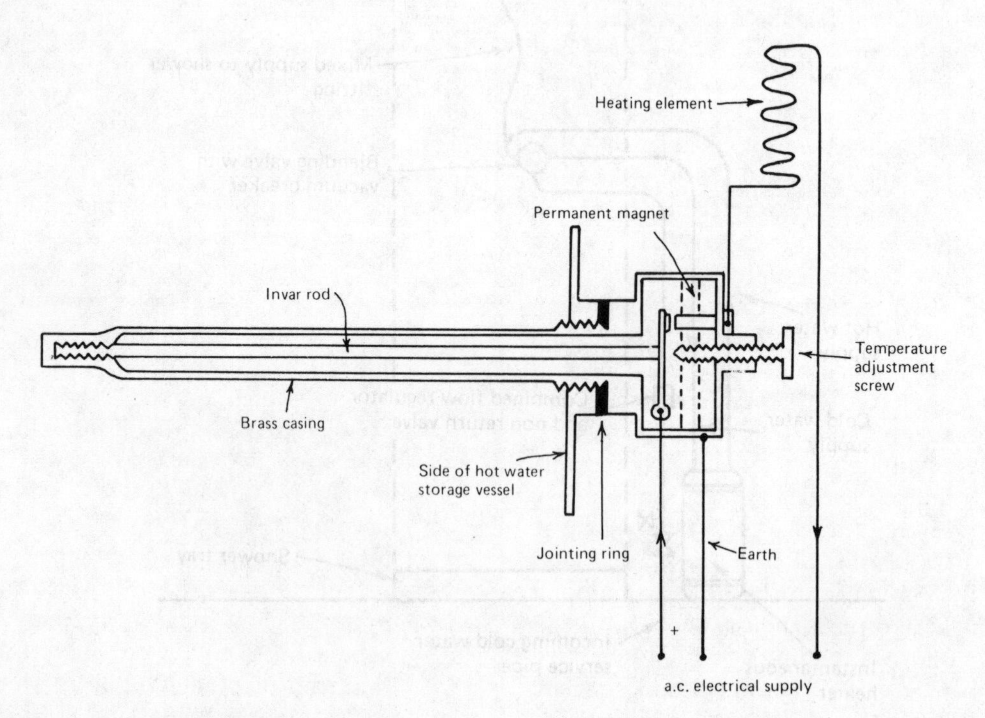

Fig. 10.10 Rod thermostat.

In breaking the circuit, the contacts are moved only a very short distance apart. This distance is often termed a 'micro-gap' and is not sufficient to extinguish the arc that is set up as the circuit is broken. The extinguishing of the arc, however, is done automatically by alternations of the supply when the voltage during the cycle reaches zero. The contacts themselves are relatively large in size and thus cool the arc to such an extent that it is prevented from restriking when the voltage reaches its maximum value again.

Gas Water Heaters

There are three main types of gas water heaters; instantaneous, storage and circulator.

Instantaneous Heaters (Fig. 10.11). These provide instant hot water and the water is heated as it passes through a heat exchanger, which is heated by the hot products of combustion. They are obtainable in the following four classes.

1. Bath heaters: these are fitted with a swivel spout outlet to serve both bath and washbasin.
2. Boiling water heaters: these are fitted with a swivel spout outlet and provide boiling water for brewing tea. A humming device fitted to the heater sounds when the water is boiling.
3. Sink heaters: these are fitted with a swivel spout outlet to provide hot water for sink use. These heaters may also be fitted over a washbasin and, if two washbasins are fitted adjacent to each other, the heater may be fitted between to serve both basins.
4. Multipoint heaters: as the name implies, these heaters may serve several fittings, such as a range of washbasins or bath, basin, sink and shower. Fig. 10.12 shows a multipoint heater for a house and Fig. 10.13 shows a multipoint heater connected to an existing hot water supply system. When the solid fuel boiler is out, hot water may be obtained from the heater by closing valve A and opening valve B.

When a hot water tap is opened, cold water passes through a venturi tube, creating a differential pressure across a flexible diaphragm. This moves the diaphragm, which is fixed to a spring-loaded valve, and allows gas to flow to the burner, where the pilot flame ignites it. The cold water flows up to the finned heat exchanger, where it is heated to the required temperature before flowing out to the open hot water tap. When the hot water tap is closed, water ceases to flow and the gas valve is closed by the action of the spring. A bi-metal flame-failure device (Chapter 2) is also fitted, which will shut off the supply of gas to the burner should the pilot flame be extinguished. The multipoint heater will deliver between 5.6 and 6.9 l/min at a temperature of about 65 °C. The head of water above the heater for operation of the valve and correct flow rate should be at least 3 m. The above description gives the basic principle of operation. Modern heaters however, have an electronic circuit which improves reliability by reducing the number of moving parts. A push button operates a spark ignition to start the boiler.

Storage Heaters (Fig. 10.14). The hot water in these heaters is stored inside a well insulated copper cylinder. There are two types.

1. Storage multipoint heater: these, like the instantaneous multipoint heater, may supply a range of washbasins, or bath, basin and sink, as shown in Fig. 10.15. Common hot water storage capacities of the heaters range from 84 to 114 l. The heater may be connected to a balanced flue.
2. Storage sink heaters: these, like the open-outlet single-point electric water heaters, are designed to serve one fitting at a time and are provided with a swivel spout outlet. They may be supplied with cold water direct from the main (depending upon the local water authority regulations) or from a

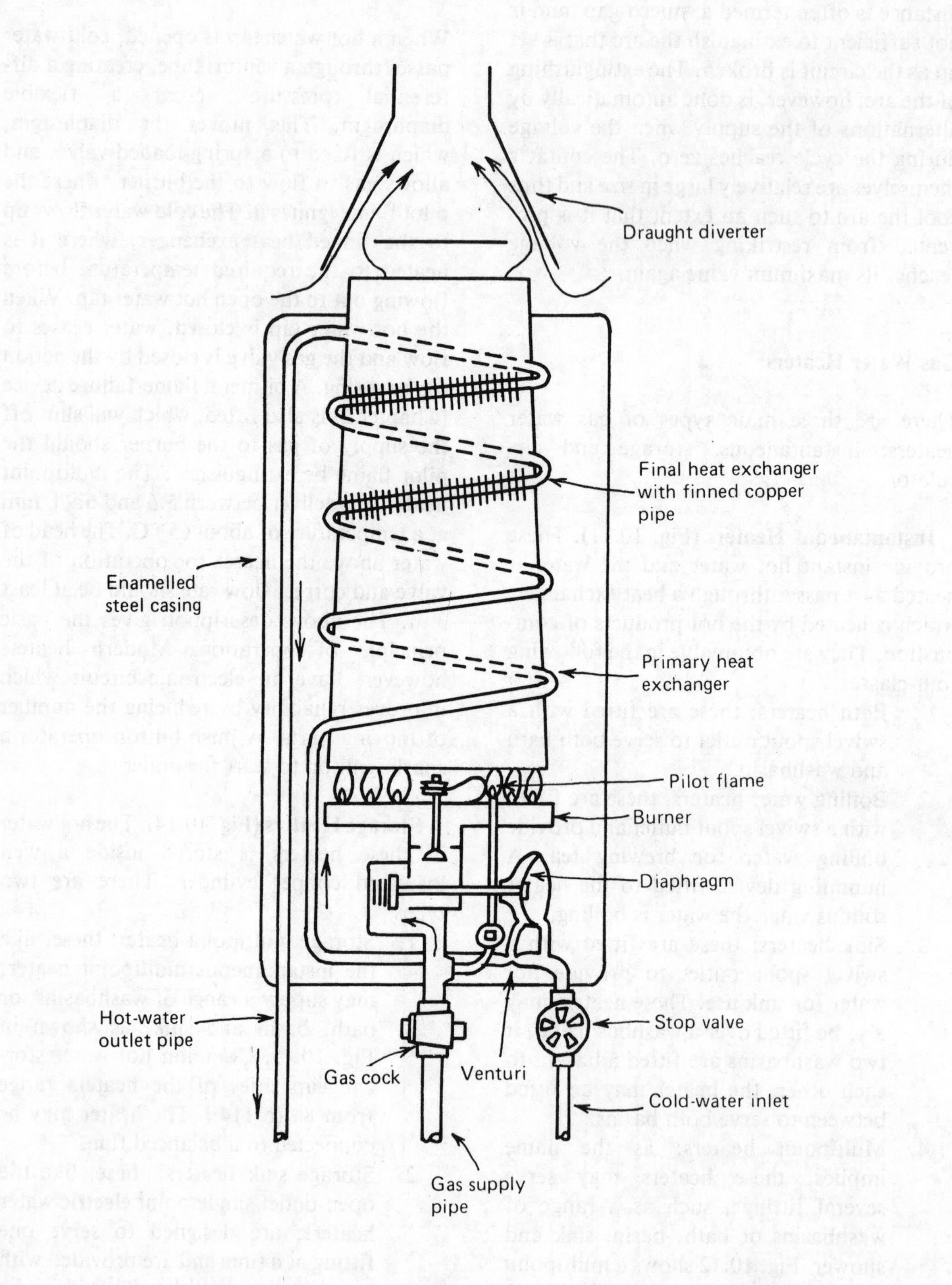

Fig. 10.11 Instantaneous gas water heater.

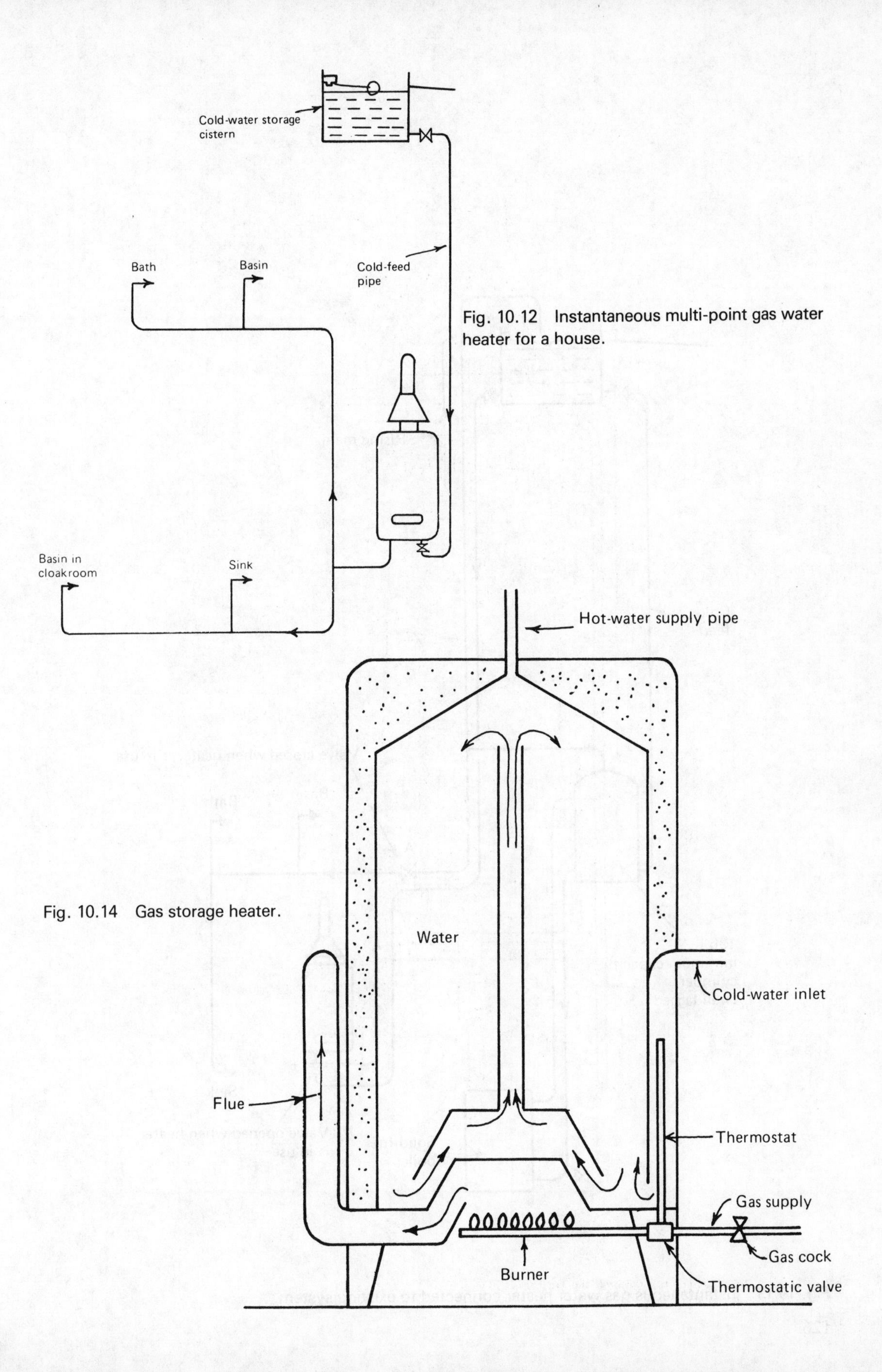

Fig. 10.12 Instantaneous multi-point gas water heater for a house.

Fig. 10.14 Gas storage heater.

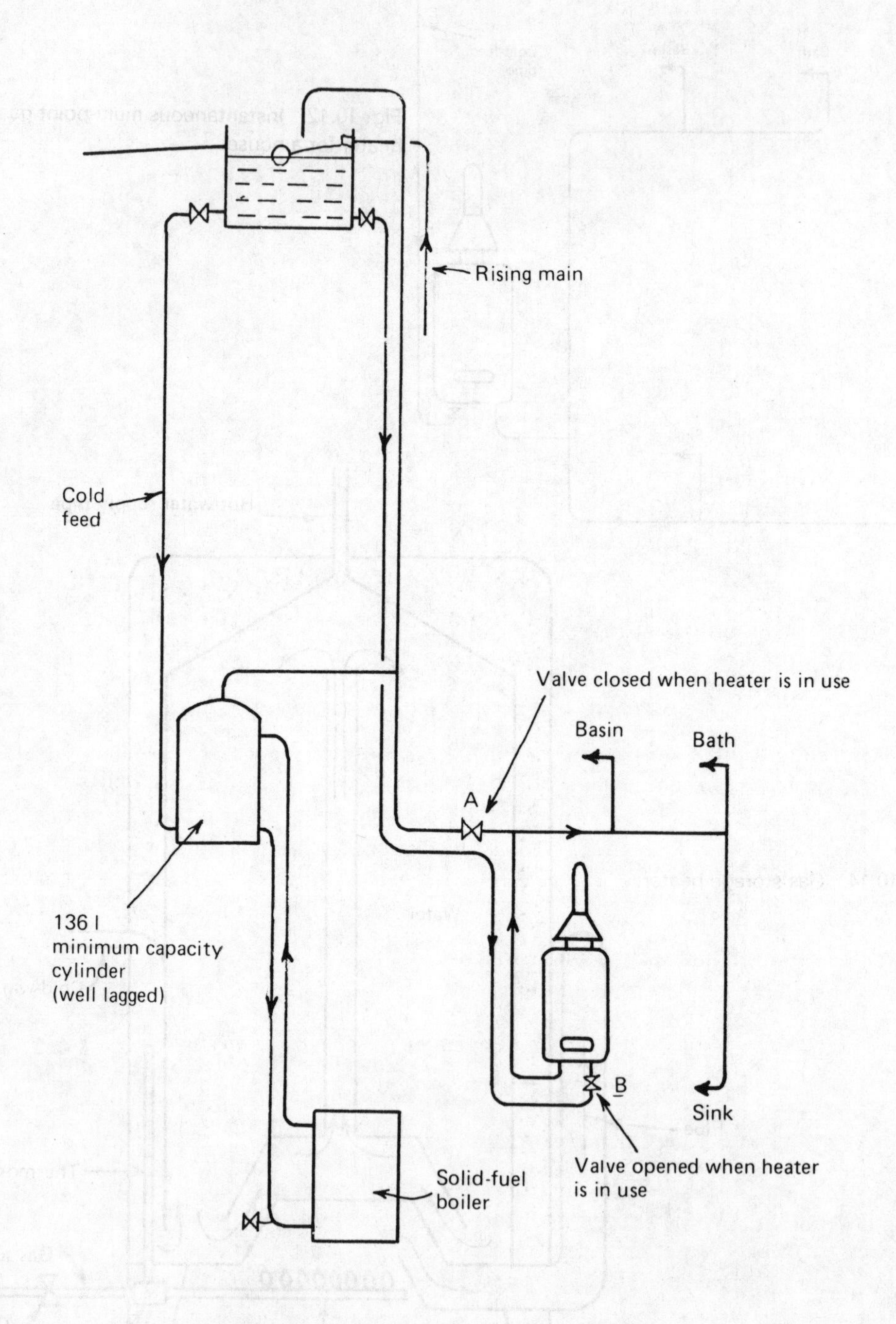

Fig. 10.13 Instantaneous gas water heater connected to existing system.

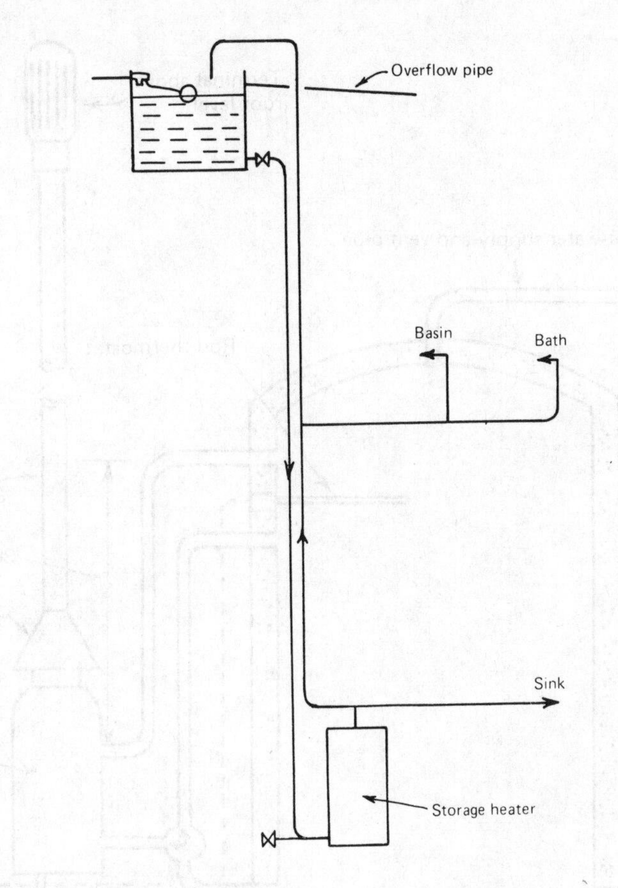

Fig. 10.15 Installation of gas storage heater.

cold water feed or storage cistern. The hot water storage capacities range from 9 to 23 l.

Note: a combination gas boiler and hot water cylinder for central heating and hot water supply has been recently brought on the market. The unit eliminates the often long primary flow and return pipes and also reduces the cost of installing heating and hot water supply. The unit comprises a 77 l capacity cylinder and an 11.7 kW boiler, which enables a 20 min hot water recovery to be achieved following a large demand (e.g. a bath).

Circulators (Fig. 10.16). These heaters are connected to the hot water storage cylinder with flow and return pipes. The heater may provide the sole means of providing hot water, or be connected to an existing hot water supply system having a solid fuel boiler and then used as an auxiliary heater (see Fig. 10.17).

When the solid fuel boiler is out, the circulator may be used in the same way as an electric immersion heater, to provide hot water. To conserve fuel, a three-way economy valve should be fitted to the return pipe. When large quantities of hot water are required, the valve can be operated so that the bottom return pipe is opened and the top return pipe closed; this allows the whole content of water inside the cylinder to be heated. When small quantities of hot water are required for sink and basin use, the valve can be operated so that the bottom return pipe is closed and the top return pipe opened; this allows only a small quantity of water at the top of the cylinder to be heated.

The usual water output of a circulator is about 25 l/h raised through a temperature of 44 °C.

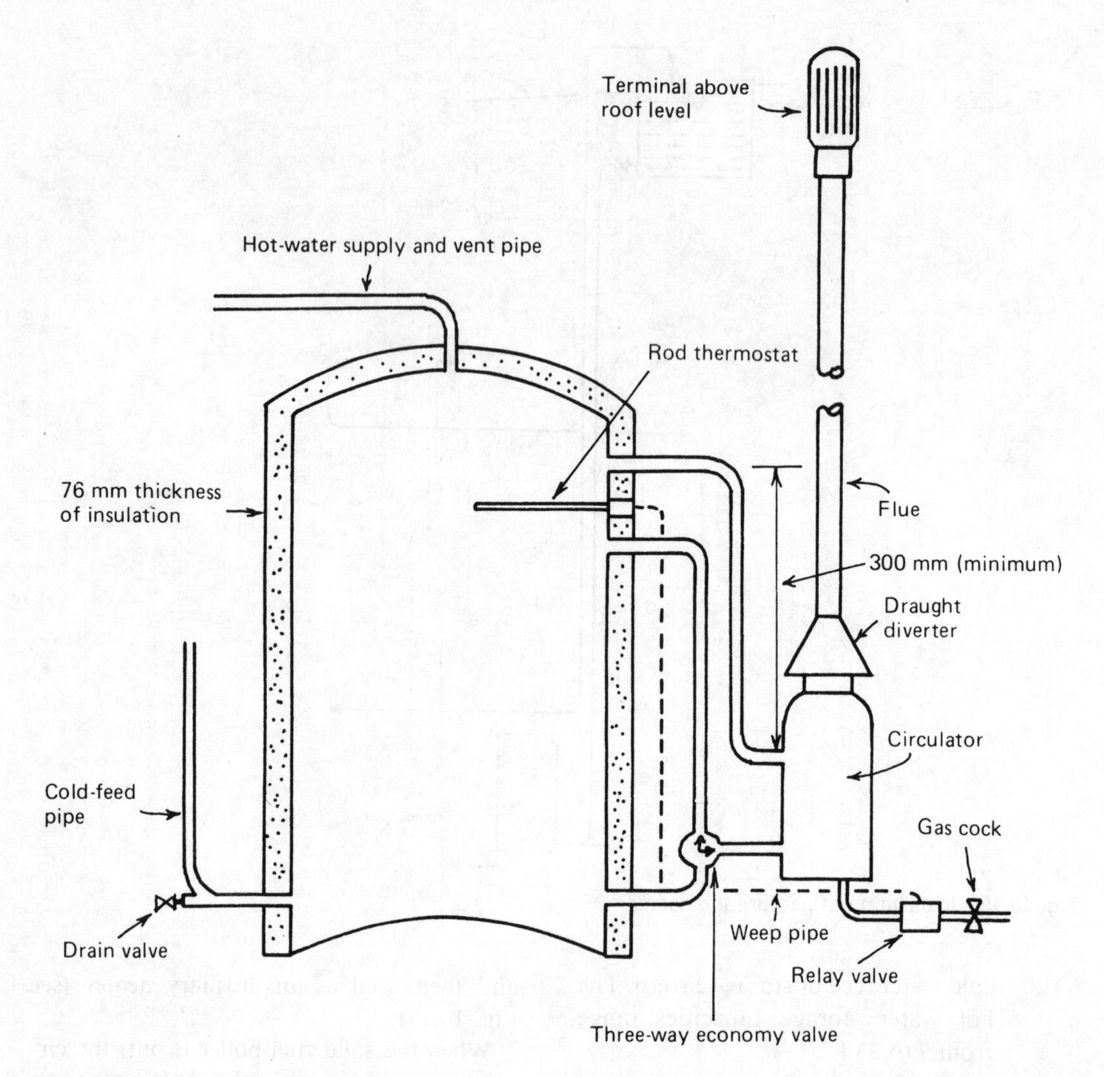

Fig. 10.16 Gas circulator.

Stratification of Water

When a storage vessel is filled with hot water and the hot water tap is opened, the hot water is forced out of the vessel by the incoming cold water. If this incoming cold water can be kept below the hot water and prevented from mixing with it, it will be possible to draw of a large proportion of the water in the vessel as hot water. The keeping of hot and cold water inside the vessel in separate layers is termed stratification.

A tall narrow cylinder is preferable to a wide short cylinder. Great care is required when cylinders are fixed horizontally, or the cold water entering may bore up through the hot water to be deflected into the upper part of the vessel. A spreader pipe is therefore required inside the cylinder connected to the cold feed pipe. Manufacturers of electric and gas water heaters provide a baffle which allows the cold water entering the heater to stratify at the base.

Exercises

1. State the merits and demerits of localised water-heating systems.
2. State the points that must be observed when installing electric water heaters in

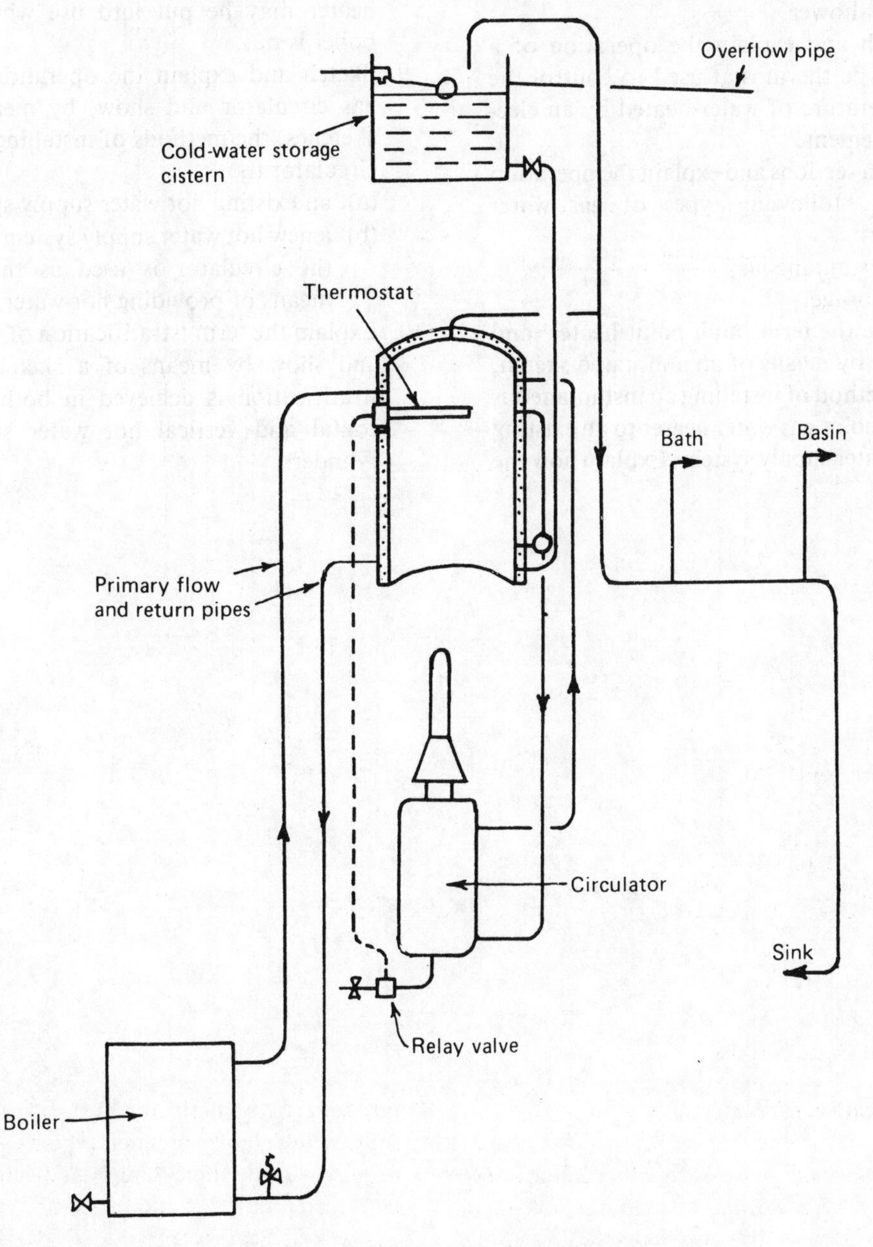

Fig. 10.17 Gas circulator connected to existing hot-water supply system.

order to reduce the heat losses to a minimum.

3. Sketch sections and explain the working principles of the following types of electric water heater:
   (a) open outlet type,
   (b) pressure type,
   (c) cistern type.
4. Show by means of annotated sketches the methods of installing pressure- and cistern-types of electric water heaters in a three-storey office to supply hot water to four washbasins on each floor.
5. Sketch a section, explain the operation of an instantaneous electric water heater and show by means of sketches the method of installing the heater to supply
   (a) a washbasin or sink,

(b) a shower.

6. Sketch and explain the operation of a rod-type thermostat used to control the temperature of water heated by an electric element.
7. Sketch sections and explain the operation of the following types of gas water heater:
   (a) instantaneous,
   (b) storage.
8. Define the term 'multipoint heater' and show, by means of an annotated sketch, the method of installing an instantaneous multipoint gas water heater to an existing hot water supply system. Explain how the heater may be put into use when the boiler is out.
9. Sketch and explain the operation of a gas circulator and show, by means of sketches, the methods of installing a gas circulator to
   (a) an existing hot water supply system,
   (b) a new hot water supply system where the circulator is used as the sole means of providing hot water.
10. Explain the term 'stratification of water' and show by means of a sketch how stratification is achieved in both horizontal and vertical hot water storage cylinders.

Chapter 11
Thermal Comfort, Low-Temperature Hot Water Heating Systems

Principles

The whole problem of achieving thermal comfort of the occupants within a building is directly related to the thermal functions of the body. If the occupants are to function at maximum efficiency, their internal temperature must be controlled within certain fixed limits. Thermal controls operate within the body itself on the rate of heat production, as well as on the rate of heat flow to the external surface of the body by the rate of the flow of blood to the surface of the skin.

There are three stages of operation of internal thermal controls of the body.

1. If the body is producing heat at a rate in excess of that being lost to the internal environment, the body perspires to increase the rate of heat loss by the latent heat of evaporation of moisture from the skin.
2. If the body loses heat in excess of the rate in which it is produced, it becomes necessary to produce more heat by physical exercise, eating more food, or wearing more clothes.
3. If the rate of heat loss from the body is 'balanced' by the heat being generated in the room, a state of thermal comfort will exist.

Thermal Comfort Conditions

The problem of providing thermal comfort is therefore one of achieving a desirable heat balance between the body and its surroundings. Fig. 11.1 shows the heat losses and heat gains of a person inside a room.

The rates of heat loss from the body and the methods by which this can be controlled are as follows.

1. Radiation: 45% (approximately) controlled by the correct mean radiant temperature of the room surfaces.
2. Convection: 30% (approximately) controlled by the correct rate of air flow past the body.
3. Evaporation: 25% (approximately) controlled by the correct rate of air flow past the body and the correct relative humidity of the air.

The greatest degree of thermal comfort exists under the following room conditions:

1. when the air temperature at feet level is not greater than 3 °C below that at head level,
2. when the air flow past the body is horizontal and at a velocity of between 0.2 and 0.25 m/s: a variable air velocity is preferable to a constant one and prevents monotony,
3. when the room surfaces are at or above the air temperature,
4. when the air temperature is between 16 °C and 22 °C, depending on the type of work being carried out and the

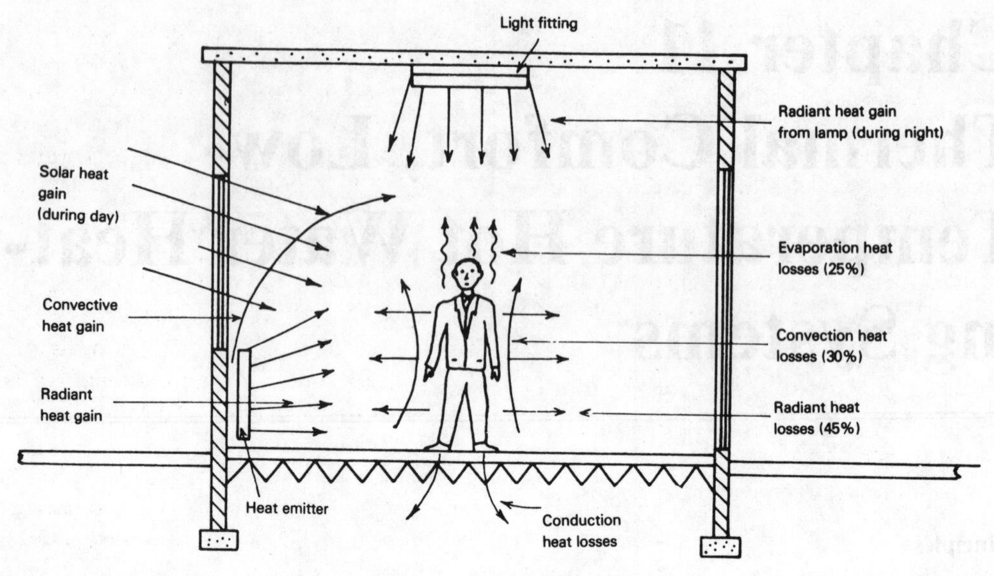

Fig. 11.1 Heat balance.

age of the occupants of the room,
5. when the relative humidity is between 40 and 60%.

Thermal Insulation

The mean radiant temperature of the external walls is an important factor in thermal comfort, and conditions are improved by good thermal insulation, which reduces heat loss from the body by radiation and of course reduces the cost of heating.

Comfort Scale

The Building Research Establishment has carried out tests on thermal comfort and produced a comfort scale similar to the one devised by Bedford; the values are established from the various tests.

Table 11.1 provides numerical values for the sensations of warmth as follows.

Table 11.1 Comfort Scale

Much too warm	7	
Too warm	6	
Comfortably warm	5	
Neither cool nor warm	4	comfort zone
Comfortably cool	3	
Tool cool	2	
Much too cool	1	

Low-Temperature Hot Water Heating Systems

Unless the system is required for the protection of stored goods, the purpose is to balance the heat losses from the occupants of the building. In the system, the temperature of the hot water is kept well below boiling point, usually at a maximum of 80 °C on the flow pipe, returning at between 10 and 20 °C below this temperature.

Although water has a high specific heat capacity of approximately 4.2 kJ/kg °C and is therefore more difficult to heat than other heat transfer media, e.g. air, 1.34 kJ/m³ °C, it can also convey more heat from the boiler to the heat emitters for the same-diameter pipe. In other words, smaller-diameter pipes may be used to convey heat by the use of water than by the use of other heat-transfer media. The higher the temperature of the water used, the smaller the diameter of pipes and heat emitters. Some of this additional heat, however, will be offset by an increase in heat losses from the pipes.

Water may be circulated in the system either by means of a pump or by natural convection. Pumped circulation has now almost replaced natural convection circulation in all but the smallest installations. It has the advantage of increasing the flow rate through the pipes, which reduces the heating-up period and allows smaller diameter pipes to be used.

Open Systems. These are provided with an expansion and feed cistern, vent pipe and cold feed pipe. Various piping circuits may be used depending on the type of building and its layout.

One-Pipe Systems. These require less pipework and reduce installation costs, but the systems have the following disadvantages over the two-pipe systems.

1. The cooler water passing out of each heat emitter flows to the next one on the circuit and this results in the heat emitters at the end of the circuit being cooler than those at the beginning of the circuit. The circuit can be balanced by partially closing the lock shield valves at the beginning of the circuit.
2. The pump forces water around the main pipe but not through the heat emitters. The heat emitters are heated by natural convection due to the difference in density between the water in the branch flow pipe and the water in the branch return pipe. The heat emitters used must therefore be of the type that offers very little resistance to the flow of water. Injector tees may be used to introduce water through the heat emitters, but these increase the resistance to the flow of water through the main cirucit.

One-pipe ring systems (Fig. 11.2) are suitable for a single-storey building and the main flow pipe may be carried overhead with the return pipe at floor level.

One-pipe drop systems (Fig. 11.3) are suitable in buildings where vertical service ducts are used and the heat emitters can be placed close to them. The system permits self venting of the heat emitters.

One-pipe ladder system (Fig. 11.4) are suitable for buildings where it is possible to run the pipes horizontal inside the room, or inside a suspended ceiling or floor duct. The heat emitters are not self venting.

One-pipe parallel systems (Fig. 11.5) are

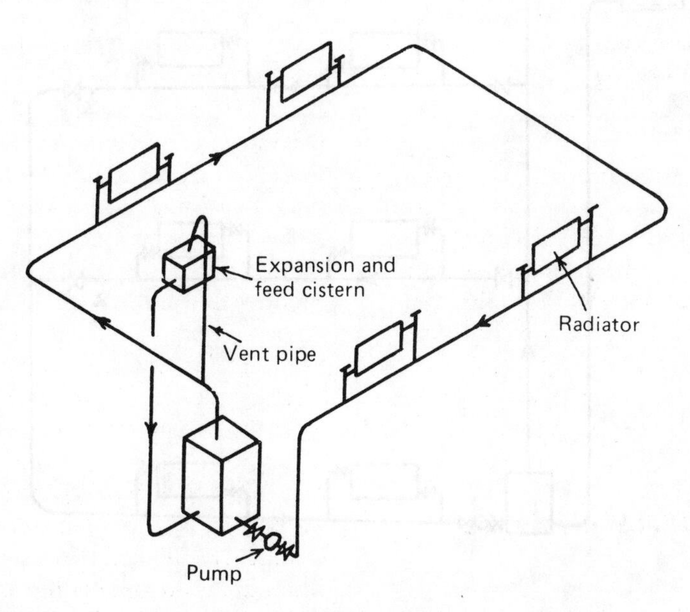

Fig. 11.2 One-pipe ring.

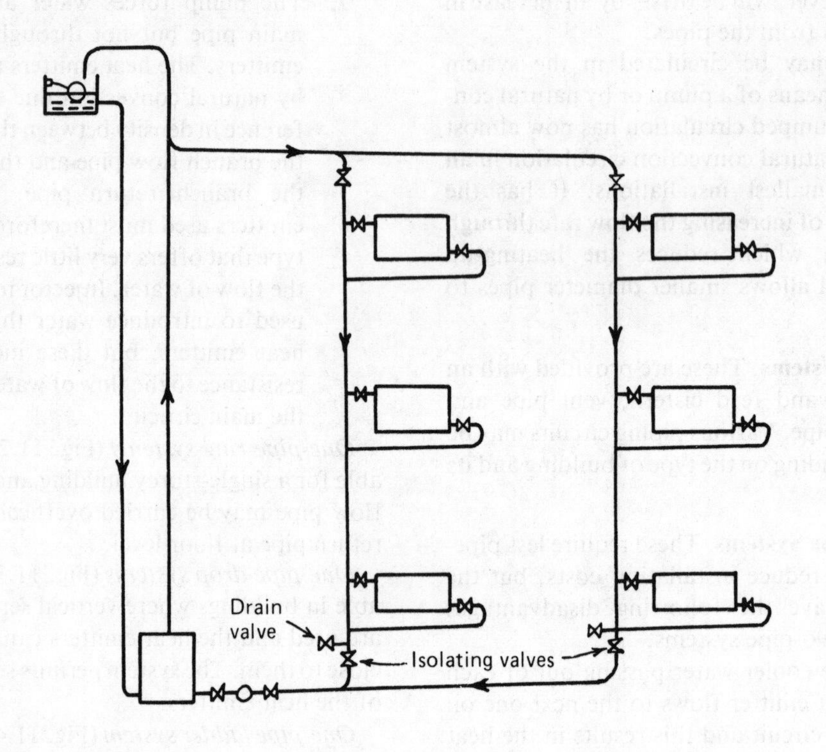

Fig. 11.3 One-pipe drop.

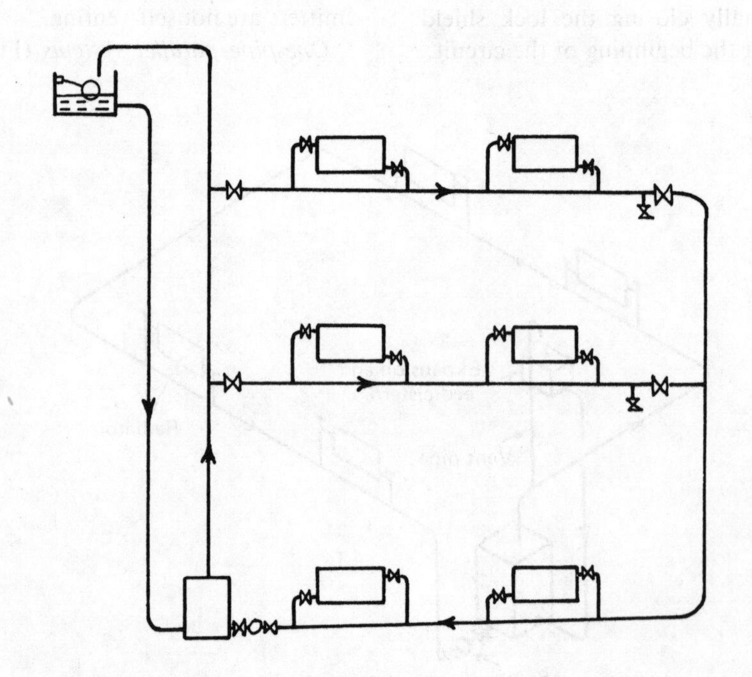

Fig. 11.4 One-pipe ladder.

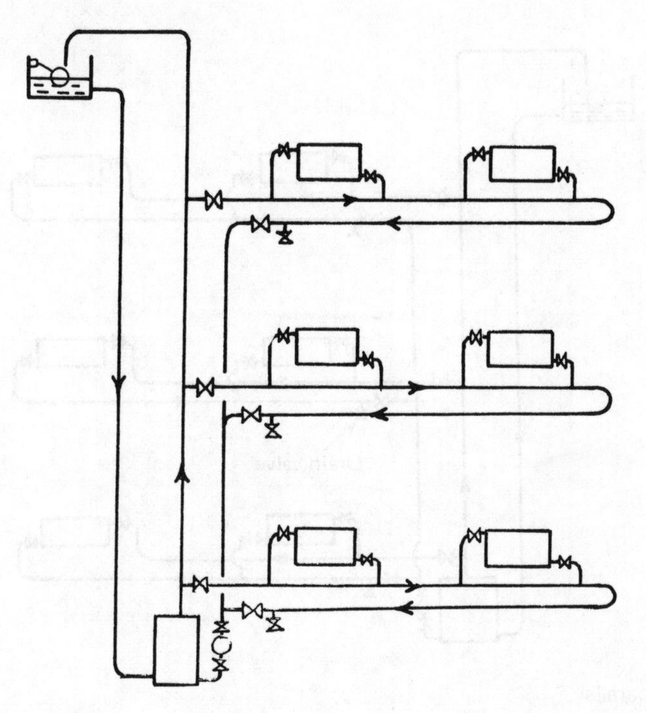

Fig. 11.5 One-pipe parallel.

similar to one-pipe ladder systems except that the return pipe is also horizontal and parallel to the flow pipe. The system is useful when it is not possible to run a vertical return pipe at the opposite end of the building to the vertical flow pipe. This may be the case when a lift shaft or a staircase is sited at the opposite end of the building to the boiler room.

Two-Pipe Systems. Although longer lengths of pipework are required for these systems, these pipes may be gradually reduced in diameter as they pass away from the boiler and gradually increased in diameter on their way back to the boiler. The cooler water from each heat emitter is carried back to the boiler by the return pipe and not passed on to the next heat emitter, as in the case of the one-pipe system. This provides a better heat balance in the system and less regulation is therefore required to even the heat distribution.

If the system is pumped, the pressure produced forces water through the heat emitters and circulation through them does not rely on natural convection, as in the case of one-pipe systems. Heat emitters offering greater resistance to the flow of water may therefore be used with the two-pipe system, e.g. convector heaters and overhead unit heaters.

Two-pipe parallel systems (Fig. 11.6) are similar to the one-pipe parallel system, but the return pipes from each emitter are connected to the main return pipe.

Two-pipe high-level return systems (Fig. 11.7) may be used when it is impractical to install a main return pipe at low level. The system is useful when installing a heating system in an existing building having a concrete ground floor and saves the cost and inconvenience of preparing a duct for the horizontal return pipe in the concrete.

Two-pipe drop systems (Fig. 11.8) are similar to the one-pipe drop, but separate flow and return mains supply the heat emitters. The system is self venting.

Two-pipe upfeed systems (Fig. 11.9) are suitable for buildings where it is impractical to install a horizontal flow main at high level. The system is not self venting and air valves are therefore required at the top of each ver-

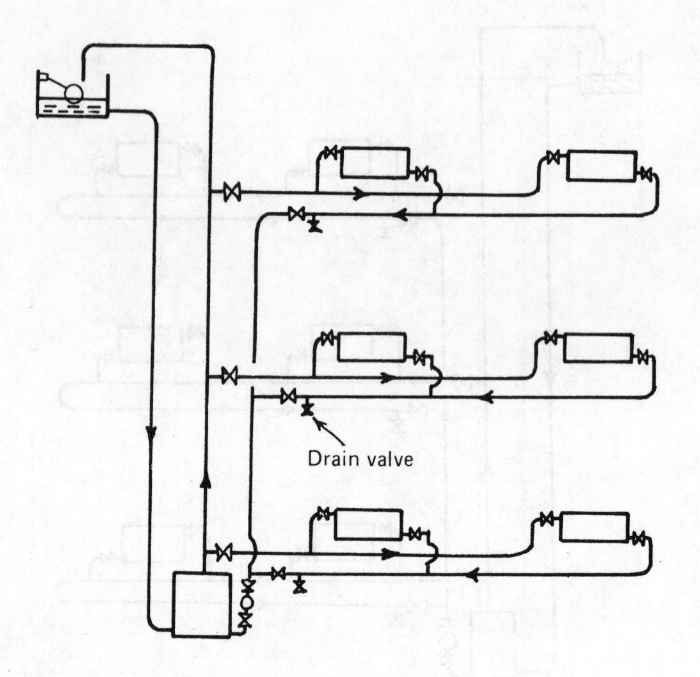

Fig. 11.6 Two-pipe parallel.

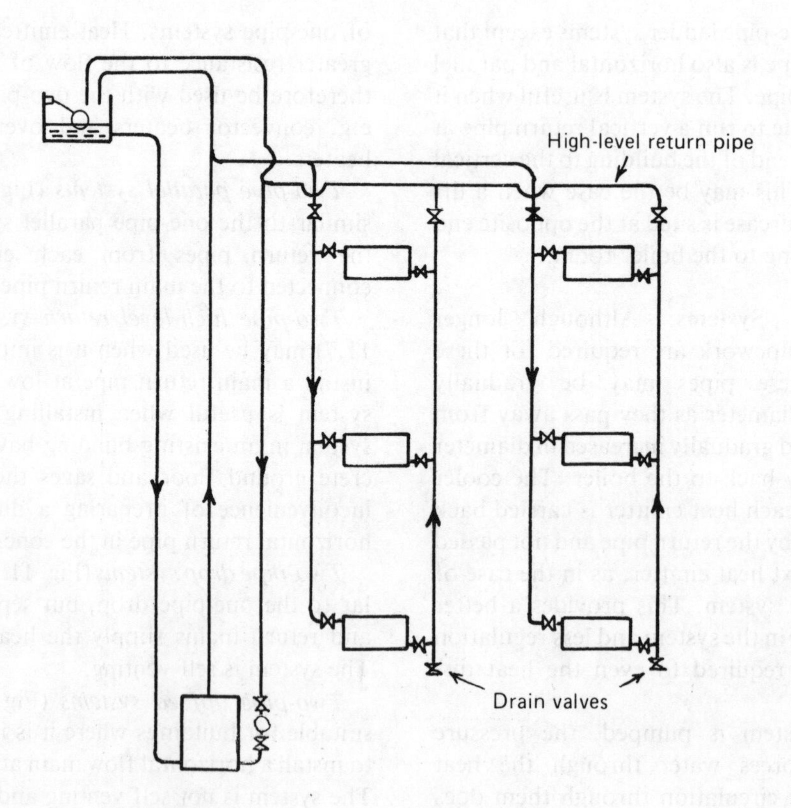

Fig. 11.7 Two-pipe high-level return.

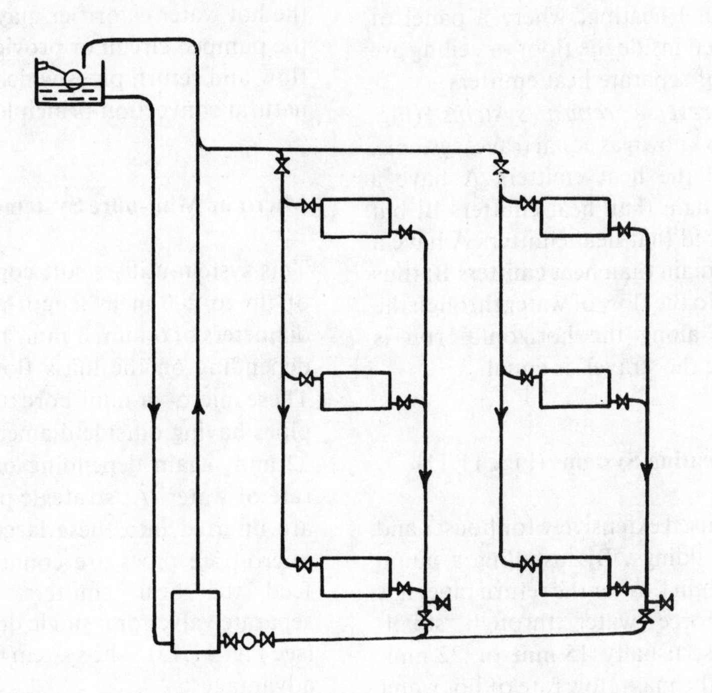

Fig. 11.8 Two-pipe drop.

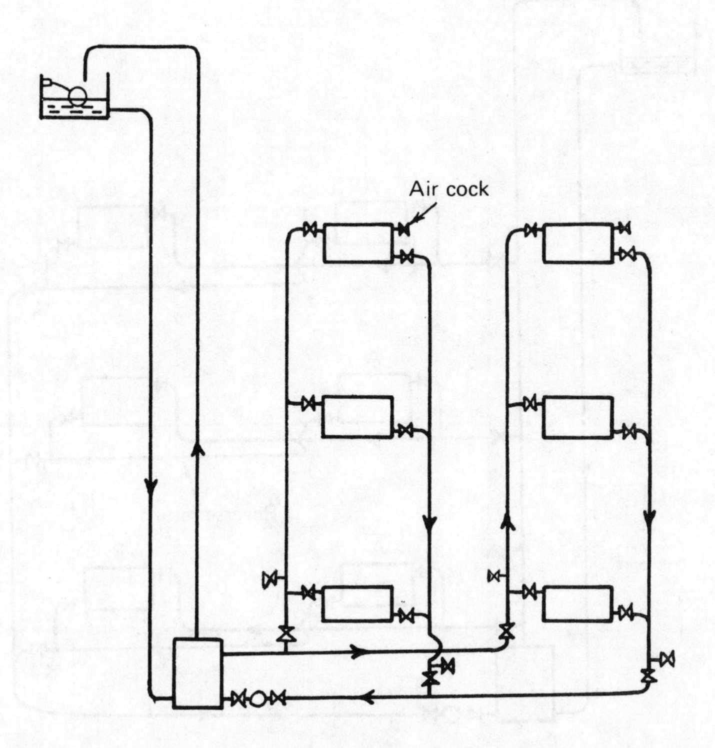

Fig. 11.9 Two-pipe upfeed.

tical flow pipe. The system is often used for embedded panel heating, where a panel of pipes embedded inside the floor or ceiling are used instead of separate heat emitters.

Two-pipe reverse return systems (Fig. 11.10) are also known as equal travel systems. In Fig. 11.10 the heat emitters A have a shorter flow main than heat emitters B, but it will be noticed that heat emitters A have a longer return main than heat emitters B: thus the resistance to the flow of water through the heat emitters along the horizontal run is equal, because the 'travel' is equal.

Small-Bore Heating Systems (Fig. 11.11)

This system is used extensively for houses and other small buildings. By installing a pump on either the main flow or the return pipe, it is possible to force water through small-diameter pipes, usually 15 mm or 22 mm, depending on the mass flow rate of hot water through the circuit.

One- or two-pipe circuits may be used and the hot water calorifier may be connected to the pumped circuit or provided with separate flow and return pipes which function on the natural convection principle.

Micro or Mini-bore Systems (Fig. 11.12)

This system utilises soft copper tube in coils of up to 200 m in length and with outside diameters of 6 mm, 8 mm, 10 mm or 12 mm, depending on the mass flow rate of water. These micro- or mini-bore tubes are fed from pipes having outside diameters of 28 mm or 22 mm, again depending on the mass flow rate of water. At strategic points, manifolds are inserted into these larger pipes and the micro-bore pipes are connected to these to feed the heat emitters, through either separate valves or a single double-entry valve (see Fig. 11.13). The system has the following advantages:

1. saving in installation time which can

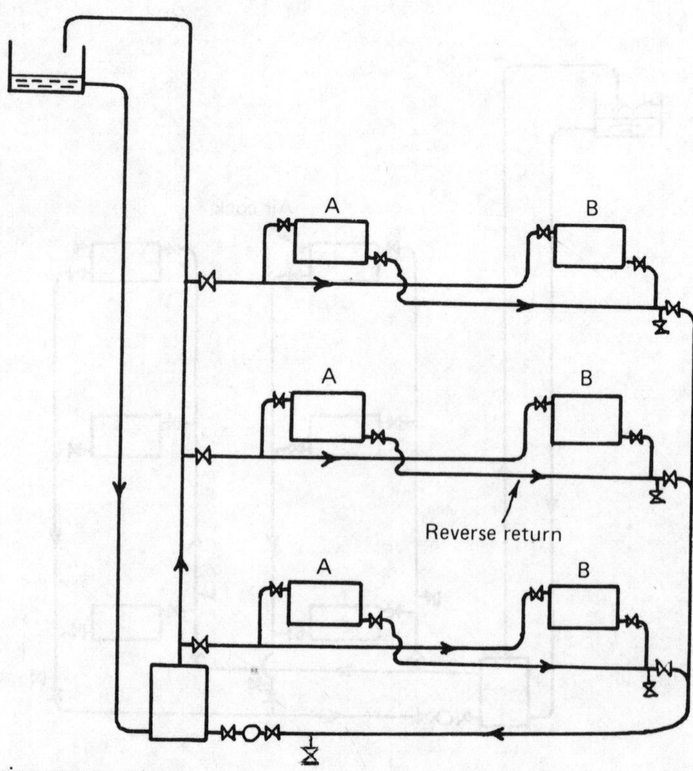

Fig. 11.10 Two-pipe reverse return.

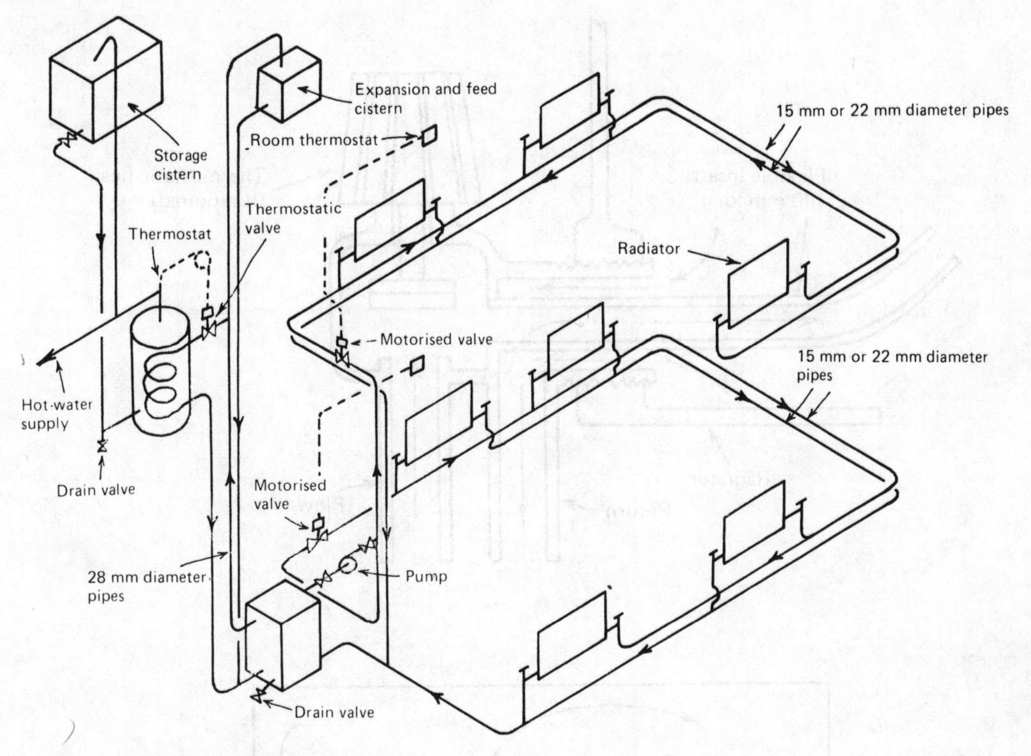

Fig. 11.11 Small-bore heating system.

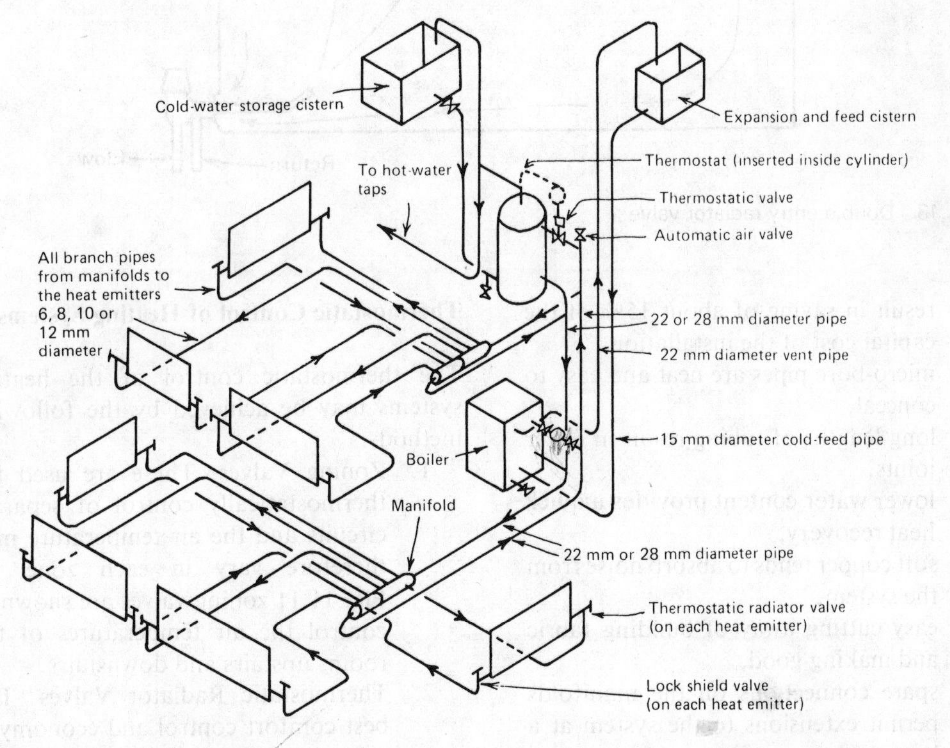

Fig. 11.12 Mini-bore or micro-bore system.

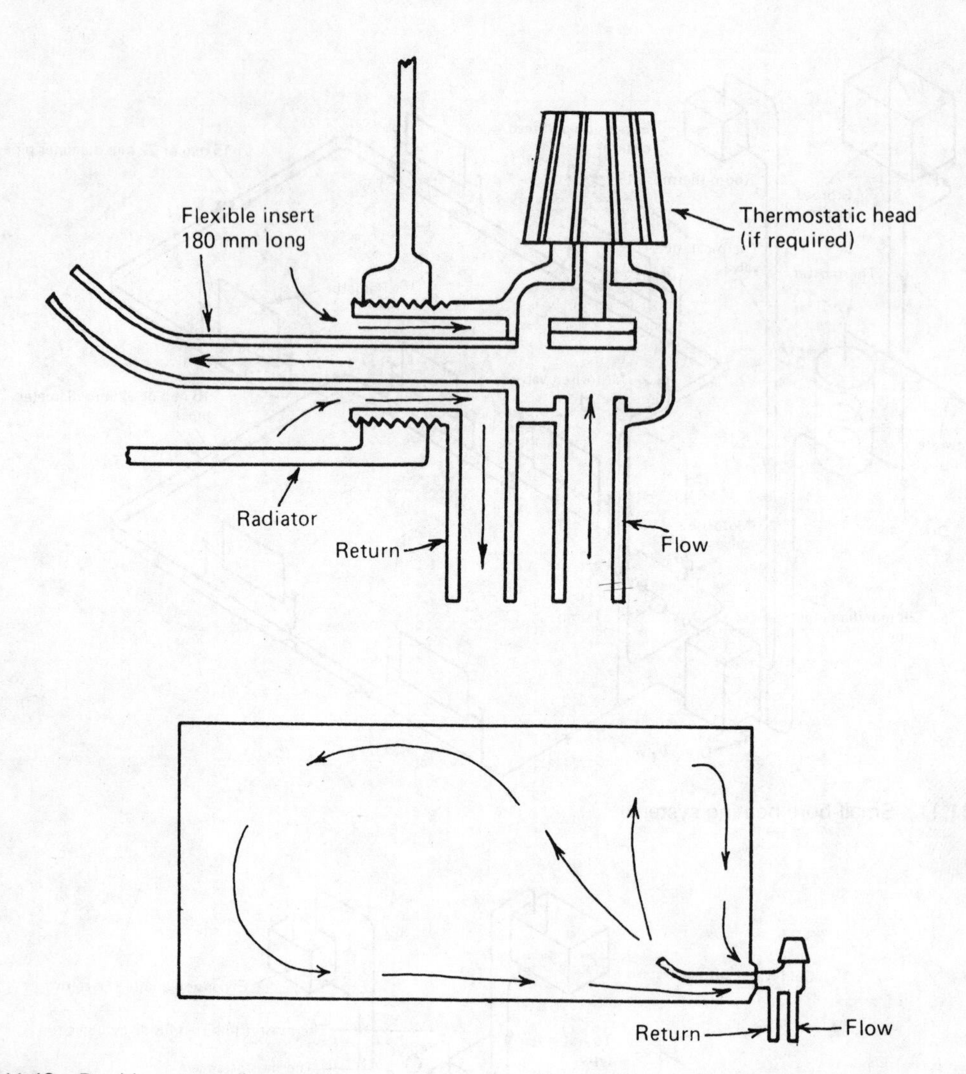

Fig. 11.13 Double entry radiator valve.

result in saving of about 15% of the capital cost of the installation,
2. micro-bore pipes are neat and easy to conceal,
3. long lengths of tubing result in fewer joints,
4. lower water content provides a quick heat recovery,
5. soft copper tends to absorb noise from the system,
6. easy cutting away of building fabric and making good,
7. spare connections on the manifolds permit extensions to the system at a later date.

Thermostatic Control of Heating Systems

The thermostatic control of the heating systems may be achieved by the following methods.

1. Zoning Valves. These are used for thermostatically control of separate circuits and the air temperature may therefore vary in each zone. In Fig. 11.11 zoning valves are shown to control the air temperatures of the rooms upstairs and downstairs.
2. Thermostatic Radiator Valves. The best comfort control and economy is achieved when each radiator is

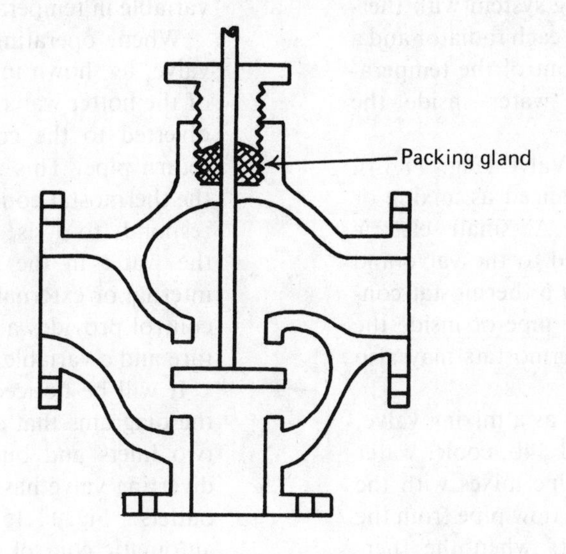

(a) Section through valve

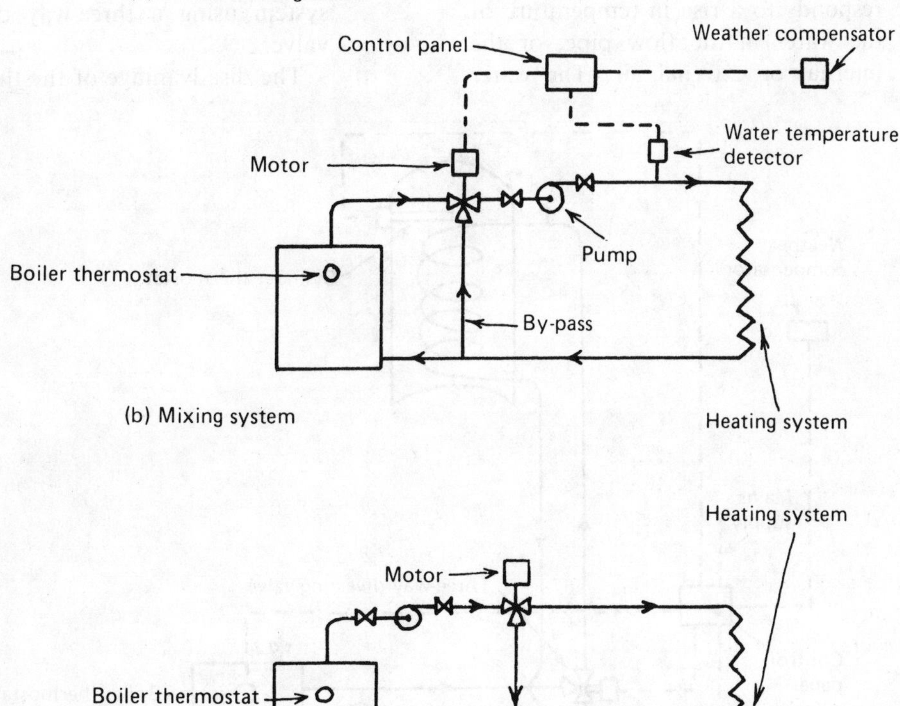

(b) Mixing system

(c) Diverting system

Fig. 11.14 Three-way valve.

equipped with a thermostatic radiator valve. Fig. 11.12 shows a mini-bore or micro-bore heating system with thermostatic valves on each radiator and a similar valve to control the temperature of the hot water inside the calorifier.

3. Three-Way Port Valve (Fig. 11.14). These may be arranged as mixing or diverting valves. A small electric motor is connected to the valve and this is operated by a thermostat connected to the flow pipe or inside the room; external thermostats may also be used.

When operating as a mixing valve, as shown in Fig. 11.14b, cooler water from the return pipe mixes with the hotter water in the flow pipe from the boiler. This occurs when the thermostat connected to the valve responds to a rise in temperature of the water in the flow pipe, or the internal or external air. The water flowing to the heating circuit is therefore constant in volume flow rate, but variable in temperature.

When operating as a diverting valve, as shown in Fig. 11.14c, some of the hotter water in the flow pipe is diverted to the cooler water in the return pipe. This again occurs when the thermostat connected to the valve responds to a rise in temperature of the water in the flow pipe, or the internal or external air. This form of control provides a constant temperature and a variable volume flow rate.

It will be noticed by inspection of the diagrams that a mixing valve has two inlets and one outlet, whilst a diverting valve has one inlet and two outlets. Fig. 11.15 shows a fully automatic control system for a combined heating and hot water supply system using a three-way diverting valve.

The disadvantage of the three-way

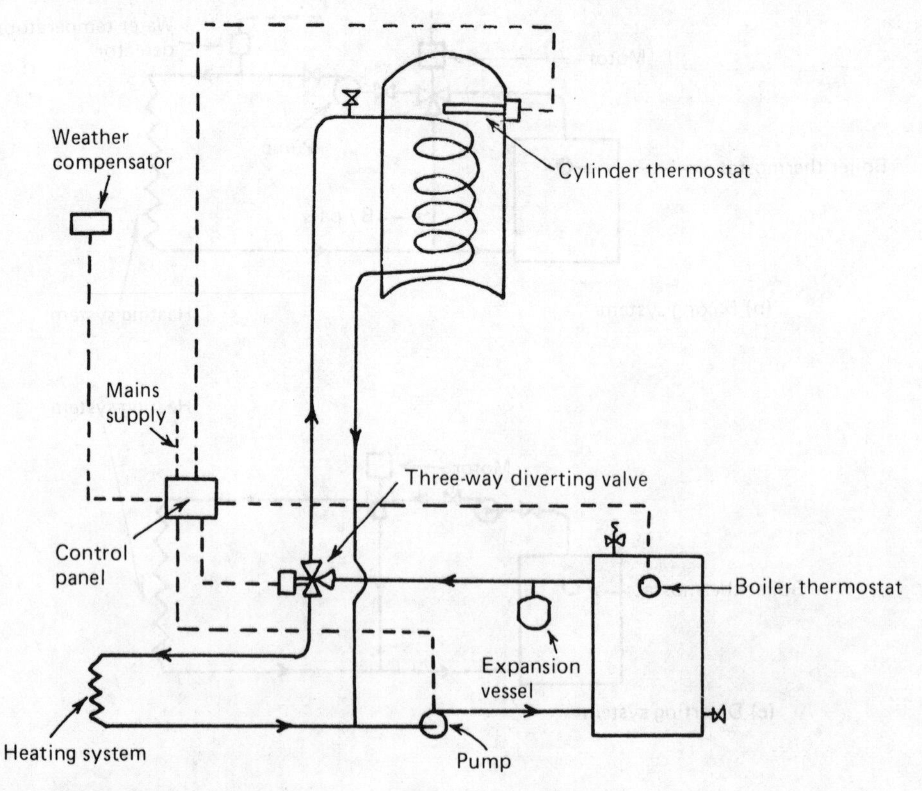

Fig. 11.15 Full automatic control.

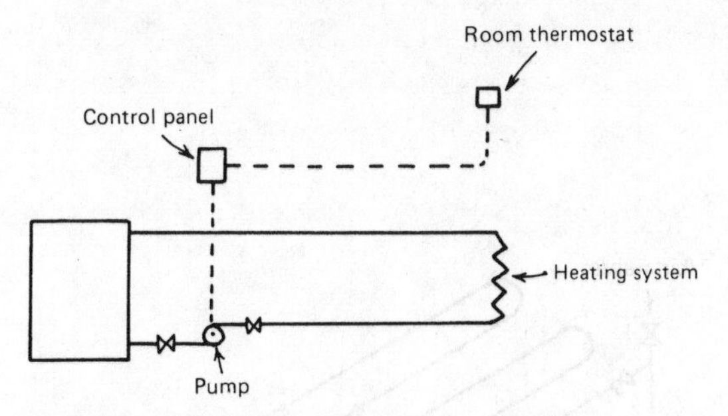

Fig. 11.16 Room thermostat controlling the pump.

port valve is in the fact that the method cannot take into account the different orientation or heat gains in the rooms. If an external thermostat sited on the north side of the building is used, any solar heat gains on the south side will not be taken into account. If an internal thermostat is used, the method cannot take into account other rooms, which may receive a significant amount of heat from occupants, lighting and machinery.

Thermostat Controlling the Pump (Fig. 11.16). This is the cheapest method of thermostatic control and is used extensively for houses. A room thermostat regulates the heat output from the heat emitters by switching the circulating pump on and off.

Cylinder Thermostat for Hot Water Priority (Fig. 11.17). This method provides a quick recovery of hot water temperature in the cylinder after a large quantity of hot water has been drawn off. Whenever the cylinder falls below the setting of the cylinder thermostat, the room thermostat is overridden and the pump switched off. Heating of the radiators is resumed as soon as the cylinder water temperature has recovered.

The two methods have the same disadvantage as the three-way port valve and cannot take into account the different orientations or heat gains in the rooms.

Boiler Control

The boiler must be provided with its own thermostatic control and the operating temperature must be sufficient to prevent condensation of the flue gases. The control of the boiler must also be under the influence of a time switch, which will allow the boiler to be started up and shut off at predetermined times. Alternatively, it may be required to reduce the temperature at night time, which requires an additional boiler thermostat, and the clock switch then changes over from the day thermostat to the night thermostat.

Embedded Panel Heating (Fig. 11.18)

As an alternative to the more conventional radiator systems, 15 mm or 22 mm diameter soft copper pipes may be embedded in the

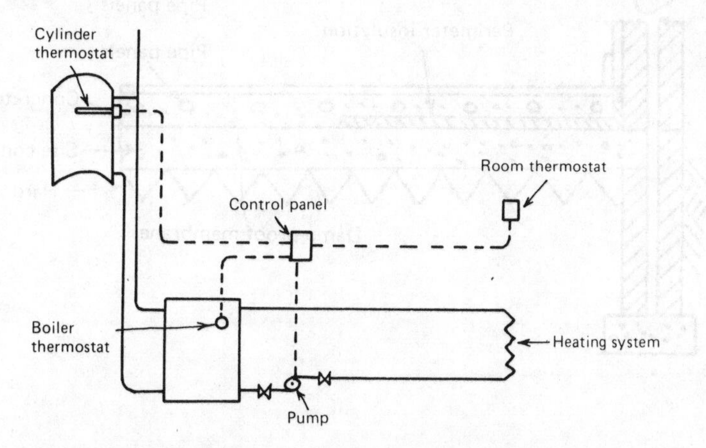

Fig. 11.17 Cylinder thermostat to give priority to the hot-water supply.

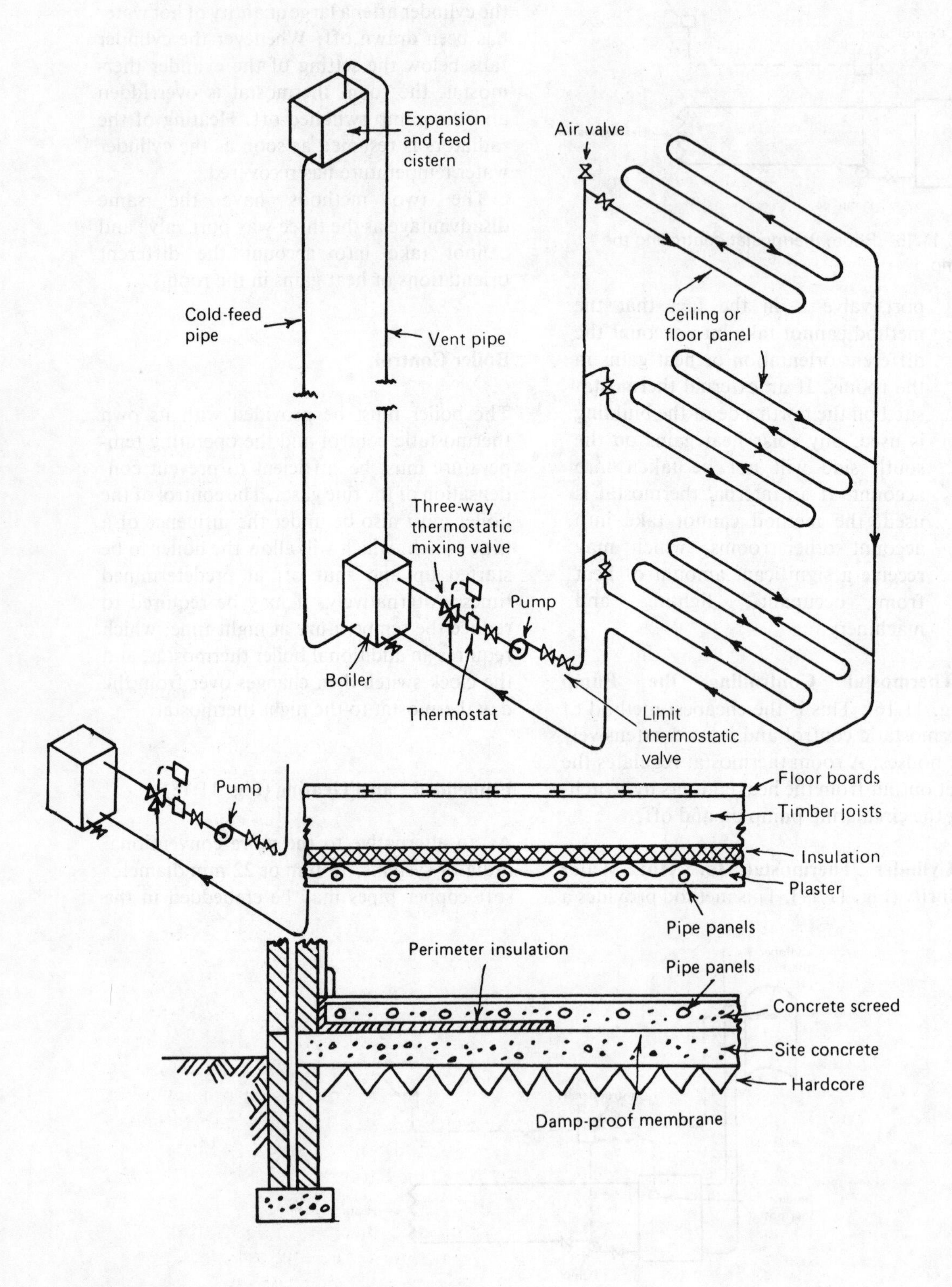

Fig. 11.18 Embedded-panel heating system.

floors, walls or ceilings. This method provides an efficient form of radiant heating and a high degree of thermal comfort.The system may be used in large buildings and houses. In the system, warm water is circulated by means of a pump through panels of closely spaced pipes at 150 mm or 225 mm centres, and a three-way thermostatic mixing valve maintains the water at an even temperature. The pipes must be thoroughly tested under water pressure before being embedded.

The temperature of the water in the flow and return pipes for different panel locations is given in Table 11.2.

Solar Space Heating

Embedded heating is particularly suitable for solar space heating where solar-heated water at a comparatively low temperature may be pumped through the panels from a large insulated underground storage tank.

De-aerator (Fig. 11.19)

Pumped 'open'-type low-pressure hot water heating systems are sometimes troublesome because of water being forced through the vent pipe or air being drawn down through it. A de-aerator recently introduced by Myson Ltd. prevents the above occurrences and also allows air from the system to escape through the vent pipe. The fitting must be connected to the flow pipe and the pipework must rise to it, so that air can escape. If the fitting is connected to the horizontal flow pipe, it should be sited as close as possible to the boiler. It must not be fitted below the level of the boiler flow pipe outlet.

Table 11.2 Flow and Return Temperatures

Panel location	Flow water (°C)	Return Water (°C)
Ceiling	54	43
Floor	43	35
Wall	43	35

Exercises

1. Explain the factors that control the thermal comfort of an occupant in a room and state the usual values required to achieve the greatest degree of thermal comfort.
2. Why does good insulation increase the thermal comfort of an occupant of a room?
3. Define the term 'comfort scale' and state its values.
4. State the advantages of using water as a heating medium for a heating system.
5. Describe the principles of operation of an 'open'-type water heating system.
6. Sketch the following pumped 'open' one-pipe heating systems:
   (a) ring,
   (b) drop,
   (c) ladder,
   (d) parallel.
7. State the advantages and disadvantages of a two-pipe system of heating.
8. Sketch the following pumped 'open' -type two-pipe heating systems:
   (a) parallel,
   (b) high-level return,
   (c) upfeed,
   (d) reverse return or equal travel,
   (e) downfeed or drop.
9. Describe the small-bore system of heating and show by means of an annotated isometric sketch how the system may be installed in a house having three radiators on the first floor and four radiators on the ground floor. Show on the sketch the method of providing hot water to a bath, basin and sink, by means of an indirect hot water system.
10. State the advantages of a micro- or mini-bore system of heating and sketch an isometric diagram of a system for a bungalow having four radiators.
11. Sketch and describe the following methods of thermostatic control of hot

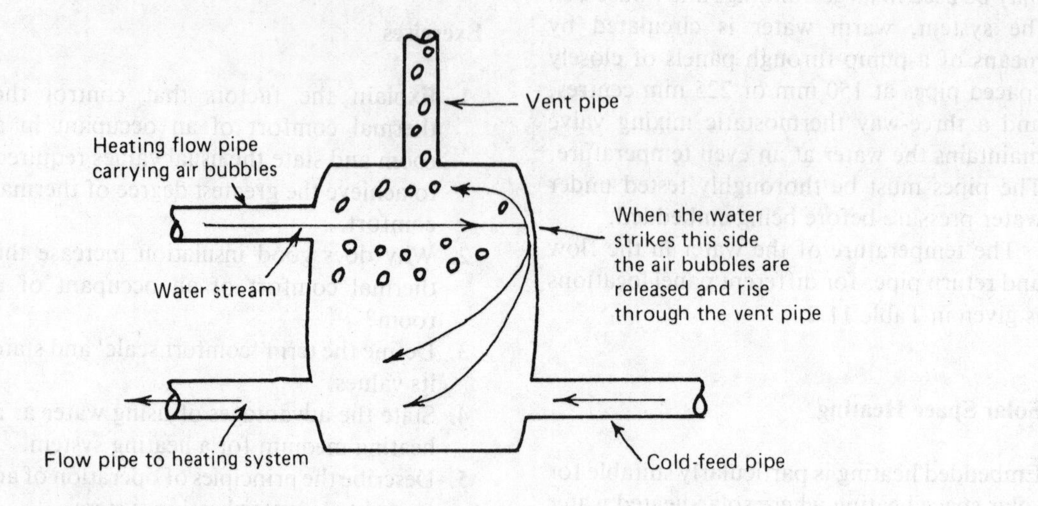

(a) Section through de-areator

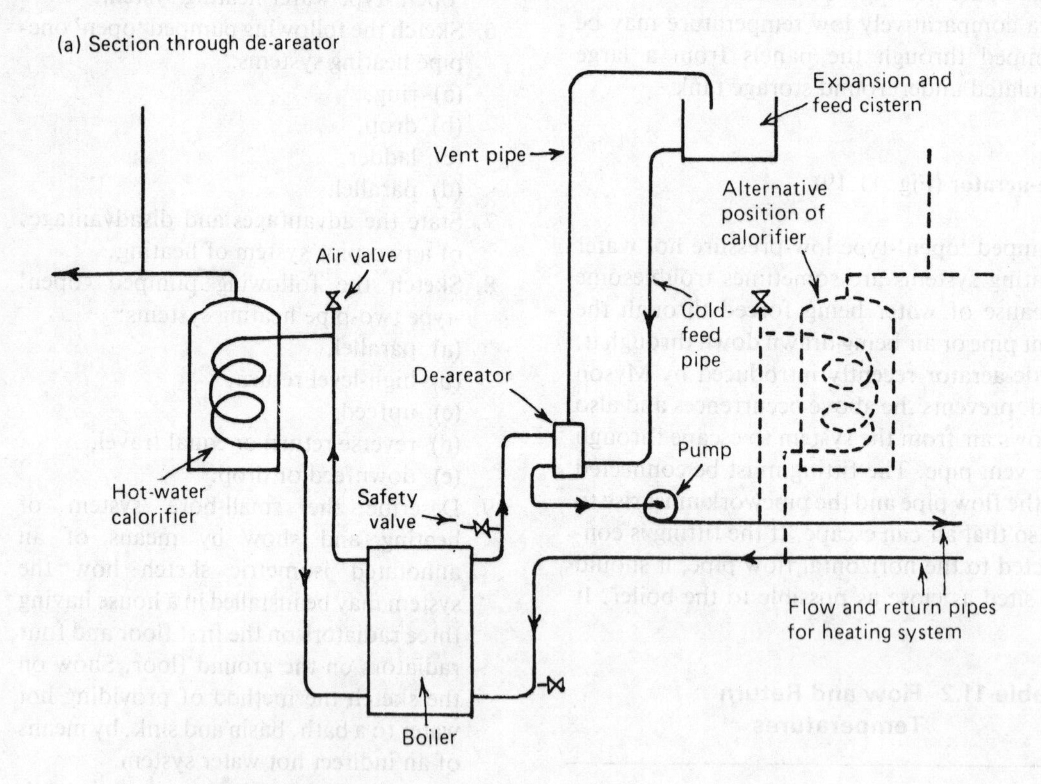

(b) Method of installing de-areator

Fig. 11.19 De-aerator.

water heating systems and state the advantages of each:

(a) thermostatic zoning valve,
(b) thermostatic radiator valve,
(c) three-way mixing valve,
(d) three-way diverting valve,
(e) switching of pump on and off.

12. Explain and illustrate by means of a sketch the method of controlling a combined heating and hot water system so that priority may be given to the supply of hot water.

13. Sketch and describe an embedded panel system of heating and sketch a section through the following structures to show the method of embedding the pipes:
(a) concrete ground floor,
(b) ceiling constructed of concrete,
(c) ceiling constructed of joists and plasterboard.

14. Show by means of a sketch and explain the operation of a de-aerator for a pumped hot water heating system.

Chapter 12
Heat Loss Calculations, Heat Emitters

Definition of Terms

Thermal Conductivity(k) (Table 12.1). In SI units, the thermal conductivity is expressed as the heat flow, in watts per square metre of surface area of a material, for a temperature difference of 1°C per metre thickness and may be expressed as

$$\frac{\text{W m}}{\text{m}^2\text{ K}}$$

The thickness divided by the area m/m² cancels to 1/m and the expression results in W/m K.

Thermal Resistivity (r). This is also a property of a material, regardless of its size or thickness. It is the reciprocal of the conductivity. The unit is

m K/W

Note: in this case m²/m cancels to m.

Thermal Resistance (R). This is a product of the thermal resistivity and thickness (L/K) and is expressed as

m² K/W

Note: if the thickness, L, is quoted in millimetres it must be converted into metres.

Thermal Transmittance (U). This is the property of a structure depending on the thickness of the material, and is a measure of its ability to transmit heat. It is defined as the quantity of heat that will flow through a unit area, in unit time, per unit difference of temperature between the inside and outside environment. It is calculated as the reciprocal of the sum of the resistances of each layer of construction and the resistances of the inner and outer surfaces and of any air space or cavity. It is expressed as

W/m² K

Table 12.1 Thermal Conductivities

Material	k value
Brickwork	
light	0.80
average	1.20
dense	1.50
Concrete	
dense	1.40
aerated	0.14
Clinker block	0.05
Glass wool	0.034
Plaster	
gypsum	0.40
vermiculite	0.20
Rendering (cement and sand)	0.53
Slates	1.50
Stone	
limestone	1.50
sandstone	1.30
Timber	
softwood	0.14
hardwood	0.16

Calculation of U Values

For a simple structure the thermal transmission is expressed as

$$U = \frac{1}{R_{si} + R_{so} + R_a + R_1 + R_2 + R_3}$$

where R_{si} is the inside surface resistance, R_{so} is the outside surface resistance, R_a is the resistance of the air space, and R_1 R_2 and R_3 are resistances of structural components. Since the thermal resistance, R, is L/k, the expression may also be given as

$$U = \frac{1}{R_{si} + R_{so} + R_a + \frac{L_1}{k_1} + \frac{L_2}{k_2} + \frac{L_3}{k_3}}$$

(see Fig. 12.1). Table 12.1 gives the thermal conductivity of some common building materials.

Surface Resistances

The standard values for the internal surface resistances are given in Table 12.2.

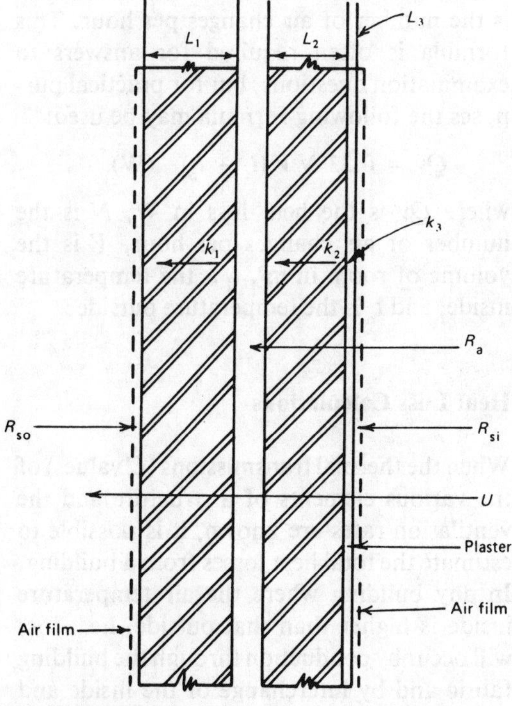

Fig. 12.1 Thermal values.

Table 12.2 Internal Surface Resistances

Internal surface resistance (R_{si})	m^2 K/W
Walls	0.12
Floors	0.15
Ceilings	0.11

Table 12.3 External Surface Resistance (R_{so}), m^2 K/W

Building element	Surface emissivity	Exposure: Sheltered	Standard	Severe
Wall	High	0.080	0.053	0.027
	Low	0.106	0.062	0.027
Roof	High	0.070	0.044	0.018
	Low	0.088	0.053	0.018

In the past, the values used for the externa. surface resistances of walls have been varied according to their orientation. The Building Research Establishment (BRE) suggest that the values should instead take into account the degree of exposure as given in Table 12.3. In column two of Table 12.3, the emissivity should be taken as high for all common building materials, including glass. Unpainted or untreated metal surfaces, such as aluminium or galvanised steel, should be taken as having low emissivity. The exposure categories should be taken as follows:

1. Sheltered. The first two storeys above ground of buildings in the interior of towns.
2. Normal. The third to the fifth storeys of buildings in the interior of towns and most suburban and country buildings.
3. Severe. The sixth and higher floors of buildings in the interior of towns and buildings exposed on hill sites, coasts and riversides.

The standard thermal resistances of unventilated air spaces are given in Table 12.4.

Example 12.1. Calculate the U value for the external wall of the house shown in Fig. 12.2.

Table 12.4 Thermal Resistances of Unvented Air Spaces

Type of air space		Thermal resistance m^2 K/W	
Thickness (mm)	Surface emissivity	Heat flow horizontal	Heat flow downwards or upwards
5	high	0.11	0.11
	low	0.18	0.18
20	high	0.18	0.21
	low	0.35	1.06

(Thermal conductivities: brickwork, 1.20 W/m K; clinker block, 0.14 W/m K, plaster, 0.40 W/m K. Thermal resistances: inside surface, 0.120 m² °C/W; outside surface, 0.053 m² °C/W; cavity, 0.180 m² °C/W.)

$$U = \frac{1}{R_{si} + R_{so} + R_a + \frac{L_1}{k_1} + \frac{L_2}{k_2} + \frac{L_3}{k_3}}$$

$$= \frac{1}{[0.12 + 0.053 + 0.18 + (0.121/1.20) + (0.08/0.14) + (0.02/0.40)]}$$

$$= \frac{1}{1.073}$$

$$= 0.93 \text{ W/m}^2\ {}^\circ\text{C}$$

Although it is very useful to calculate the U values of a structure, and this may be necessary in certain cases, their values are often given by various authorities. Table 12.5 gives a list of U values, which is part of a table in the Institute of Plumbing Guide Book.

Heat Loss due to Infiltration of Air

To find the heat loss by ventilation or infiltration of air, the following formula may be used:

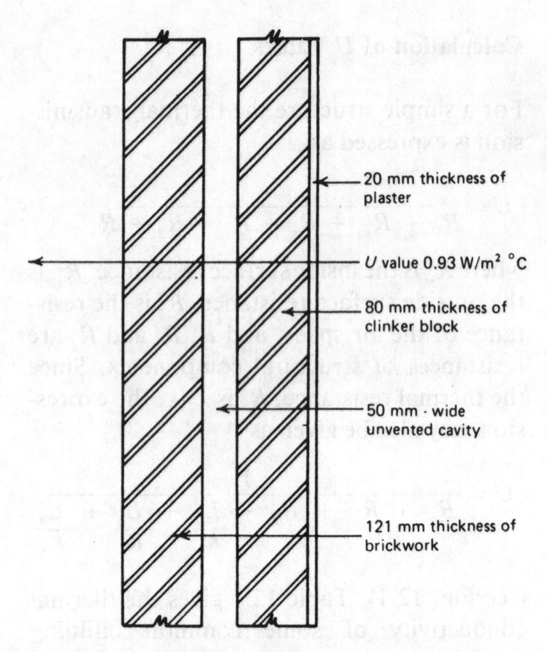

Fig. 12.2 U value for cavity wall. Example 12.1.

$$Qv = \frac{HC \times V \times (t_i - t_o) \times N}{3600} \quad \text{(W)}$$

where Qv is the heat loss in W. HC is the heat capacity of air (1340 J/m³ K), V is the volume of room in m³, t_i is the air temperature inside, t_o is the air temperature outside, and N is the number of air changes per hour. This formula is often required for answers to examination questions, but for practical purposes the following formula may be used:

$$Qv = 0.33\, N\, V\, (t_i - t_o) \quad \text{(W)}$$

where Qv is the heat loss in W, N is the number of air changes per hour, V is the volume of room in m³, t_i is the temperature inside, and t_o is the temperature outside.

Heat Loss Calculations

When the thermal transmissions (U values) of the various elements of a structure and the ventilation rates are known, it is possible to estimate the total heat losses from a building. In any building where the air temperature inside is higher than that outside, heat loss will occur by conduction through the building fabric and by interchange of the inside and outside air. These two forms of heat loss may

Table 12.5 Heat Losses through Building Fabric (U Values), W/m^2 K

Construction	Average	Exposed
External Walls		
Brickwork, solid, plastered 121 mm	3.00	3.30
Brickwork, solid, plastered 236 mm	2.10	2.30
Cavity Walls, plastered		
Both leaves 121 mm brick	1.50	1.65
Outer leaf 121 mm brick, inner leaf 104 mm insulated concrete	0.96	1.06
Windows		
Single-glazing, wood frames	4.30	4.73
Double-glazing, wood frames	2.80	2.75
Single-glazing, metal frames	5.60	6.16
Double-glazing, metal frames	3.20	3.52
Partition Walls		
121 mm brick, plastered both sides	2.30	–
75 mm breeze, plastered both sides	2.25	–
Flat Roofs		
3-layer felt on strawboard joists and plasterboard	1.05	1.08
Asphalt on 104 mm concrete and plaster ceiling	3.41	3.75
Asphalt on 150 mm concrete and plaster ceiling	3.12	3.48
Pitched Roofs		
Tiles on battens and felt	2.75	3.02
Plaster ceiling, roof space above with 75 mm glass fibre between joists	0.88	0.97
Ground Floors		
Ventilated wood floors on joists	0.61	–
Solid floors	0.36	–

be referred to as fabric heat loss and ventilation heat loss. To maintain any required temperature conditions inside the building, the rate of heat imput must be equal to the rate of heat loss. The fabric heat losses may be estimated from the following equation:

$$\text{Heat loss} = A \times U \times (t_i - t_o) \quad \text{(W)}$$

where A is the area of each element of the structure in m^2, U is the thermal transmission through the fabric in W/m^2 K, t_i is the temperature of the air inside, in °C, and t_o is the temperature of the air outside, in °C. An example will show how these heat losses are estimated.

Example 12.2 Figs 12.3 and 12.4 show the ground floor and first floor plans of a small detached house. Estimate the heat losses from the lounge and bedroom one, assuming that the outside air temperature is −1 °C, there are two air changes per hour, and the rooms are 2.4 m high and have windows sized 1.7 m × 1.4 m. Use the following U values (in W/m^2 K)

Lounge	
ground floor	0.36
external walls	0.60
internal wall	2.25
door	2.30
windows (double glazed)	2.80
ceiling	0.35
Bedroom 1	
external walls	0.60
window (double glazed)	2.80
ceiling	0.35

Note: in the lounge there will be no heat loss into the dining room, so this inner wall will be ignored. In the bedroom there will be a heat gain through the floor from the lounge; and there will be no heat loss through the inner walls so these can be ignored.

An estimate of the heat loss through the fabric can now be made (Tables 12.6 and 12.7).

Table 12.6 Heat Loss through Fabric of Lounge

Type of fabric	Area (m^2)	*U* Value (Wm^2K)	Temperature difference (°C)	Total heat loss (W)
Ground floor	18	0.36	22	142.56
External walls (less windows)	(20.4–4.76) = 15.64	0.60	22	206.45
Internal wall (less door)	(10.8–1.52) = 9.28	2.25	3	62.64
Windows	4.76	2.80	22	293.22
Door	1.52	2.30	3	10.49
Ceiling	18	0.35	5	31.50
				746.86

Ventilation loss $Qv = 0.33\ Nv\ (t_1 - t_0)$
$= 0.33 \times 2 \times 43.2 \times 22$
$= 627.264$
Total heat loss $= 627.264 + 746.86$
$= 1374.124$ W

Table 12.7 Heat Loss through Fabric of Bedroom 1

Type of fabric	Area (m^2)	*U* Valve (Wm^2 K)	Temperature difference (°C)	Total heat loss (W)
External walls (less windows)	(20.4–4.76) = 15.6	0.60	17	159.120
Windows	4.76	2.80	17	226.576
Ceiling	18	0.35	17	107.100
				492.796
Less heat gain through floor	18	0.35	5	31.500
				461.296

Ventilation loss $Qv = 0.33\ Nv\ (t_1 - t_0)$
$= 0.33 \times 2 \times 43.2 \times 17$
$= 484.704$
Total heat loss $= 484.704 + 461.296$
$= 946$ W

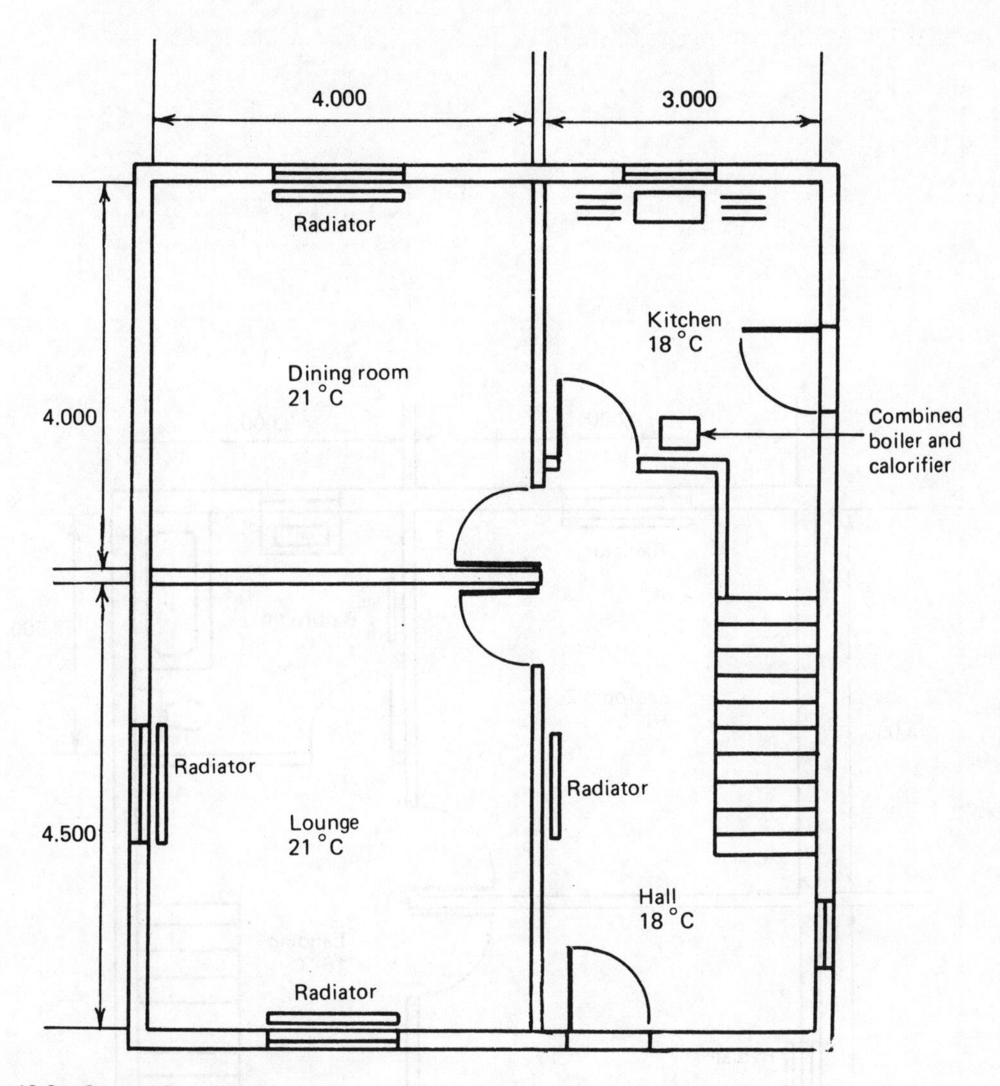

Fig. 12.3 Ground floor plan

Area of Radiators

Allowing for a heat emission from the radiator of 520 W/m² in the lounge and 530 W/m² in bedroom one, the areas of the radiators will be as follows. For lounge,

$$\text{Area} = \frac{1374.124}{520}$$

$$= 2.643 \text{ m}^2$$

Two radiators, each 1.32 m² in area, will be required and these will be fixed below the windows. For bedroom 1,

$$\text{Area} = \frac{946}{530}$$

$$= 1.785 \text{ m}^2$$

Two radiators, each 0.9 m² in area, will be required and these again will be fixed below the windows.

Note: the higher temperature difference between the air and the surface of the radiator in bedroom one will result in a higher heat emission, hence 530 W/m² instead of 520 W/m². The student should calculate the radiator areas for the other rooms.

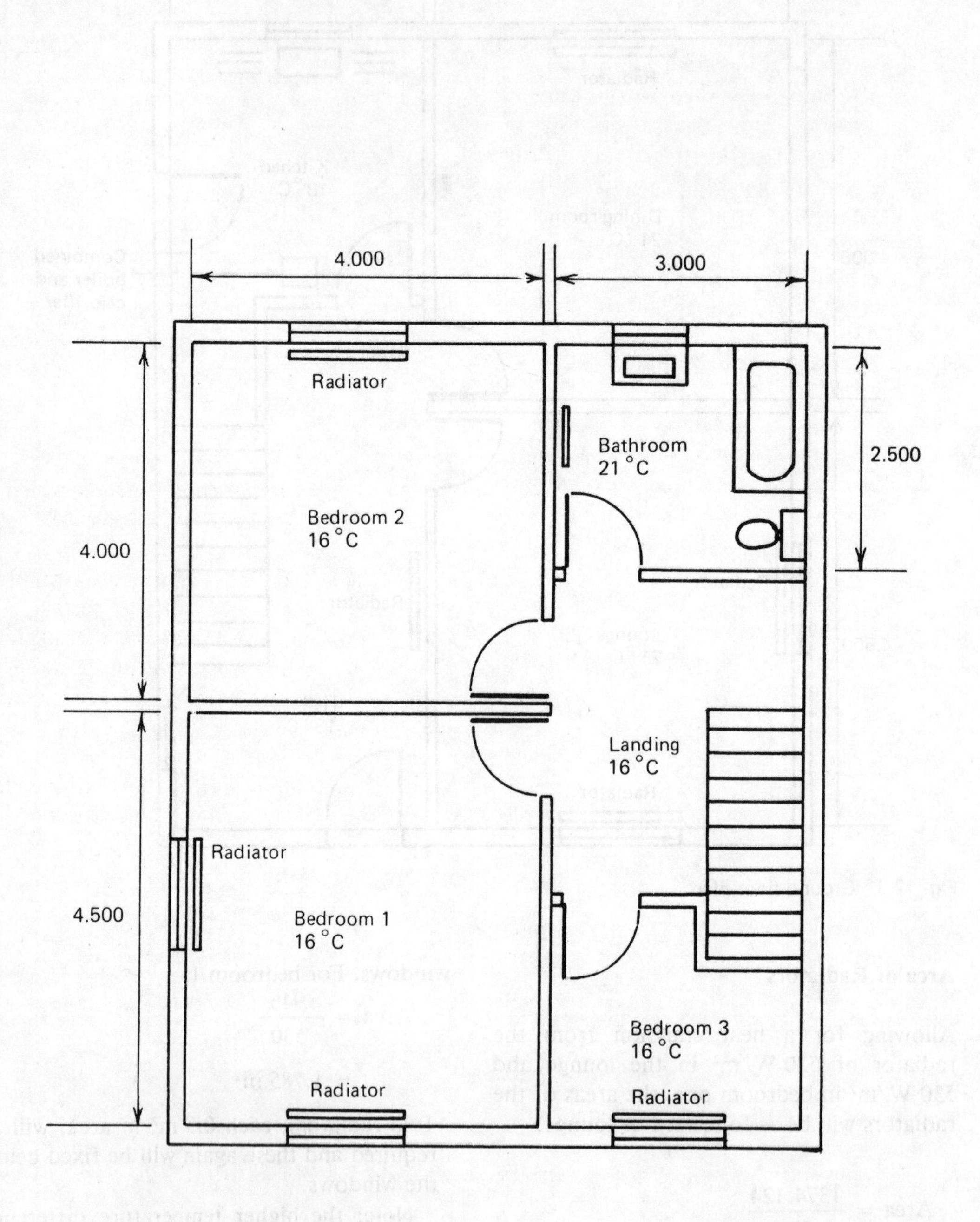

Fig. 12.4 First floor plan.

Building Regulations 1985 (Conservation of fuel and power)

The requirements of dwellings are as follows:

1. The calculated rate of heat loss (WK) through any windows and roof lights shall be no greater than it would be if:
   (a) the aggregate of the areas of windows and roof lights were 12 per cent of the area of walls bounding the dwelling and
   (b) the windows and roof lights had a *U* value of 5.7.
2. The calculated rate of heat loss through the solid parts of the exposed elements shall not be greater than it would be if:
   (a) the exposed walls and exposed floors had a *u* value of 0.6 and
   (b) the roof had a *U* value of 0.35.
3. The extent that the calculated rate of heat loss through the solid parts of exposed elements is less than the maximum permitted under sub-paragraph (2) the calculated rate of heat loss through windows and roof lights may be greater than the maximum permitted under paragraph (1).

Boiler Power

The sum of the heat losses from the various rooms will give the boiler power required in watts for the heating system and to this must be added the power required to heat the domestic hot water. A 10% margin may then be added to this total to allow for exceptionally cold weather.

Heat Emitters

Various types of heat emitter are available to suit different conditions of heating. Some types operate on convection and others on radiation, or a combination of both.

Convection. Convected heat directly warms the surrounding air and is distributed throughout the room by convection currents. Convection currents move in an upward direction, being displaced by the cooler denser air. Convected heat has the advantage of carrying heat around obstacles, which would limit the transmission of radiant heat.

Radiation. Radiant heat is transmitted from a heat emitter in straight lines and passes through the air in the form of heat waves. It is only when these waves strike an opaque surface that they are felt as heat. Radiant heat is directional, which means that a radiant heater can be placed so that it radiates heat to selected areas of a room. This heat, however, can be interrupted by an opaque body placed in its path.

Radiant heat has the following advantages:

1. The walls and contents of a room are kept relatively warm and the heat loss from the occupants is consequently lower.
2. The air temperature is lower which is more conducive to physical and mental activity.
3. The lower air temperature results in about 15% saving in fuel costs.
4. Draughts are reduced to a minimum.
5. When radiant heat warms the surfaces on which they strike, some convected heat is transmitted from them.

Column Radiators (Fig. 12.5). Most of the heat from this type of emitter is transmitted

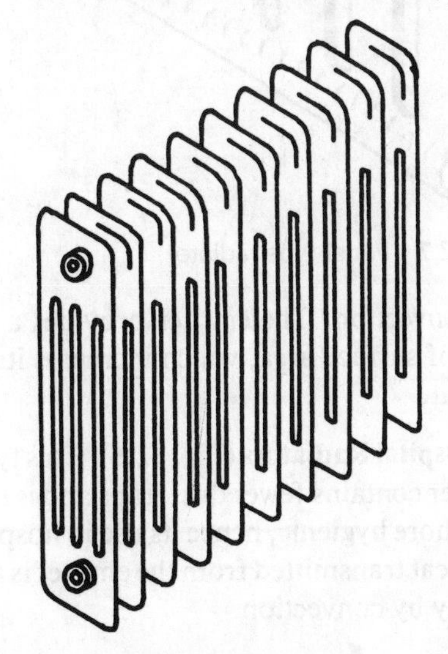

Fig. 12.5 Column-type radiator.

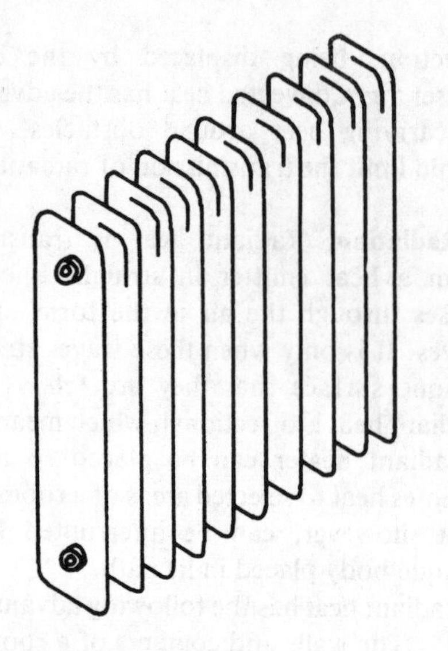

Fig. 12.6 Hospital-pattern radiator.

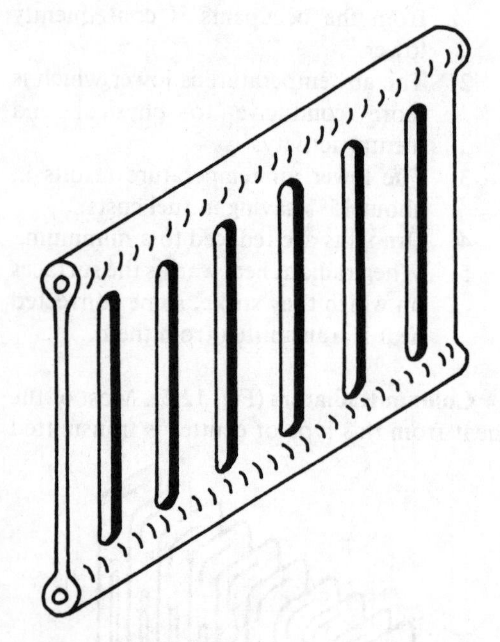

Fig. 12.7 Panel-type radiator.

by convection. The emitter provides a great deal of surface area, which increases its heat output.

Hospital Radiators (Fig. 12.6). This type of emitter contains fewer dust traps and is therefore more hygienic, hence its use in hospitals. The heat transmitted from the emitter is again mainly by convection.

Panel Radiators (Fig. 12.7). Because of their greater plane surface area, these emitters transmit most of the heat by radiation. Pressed steel panel radiators are extremely popular for domestic heating systems.

Skirting Convectors (Fig. 12.8). These consist of a continuous finned heating pipe enclosed inside a sheet metal casing. They provide excellent heat distribution by convection. When installing these heat emitters, care must be taken to ensure that the gap at the base is not closed by the carpet, or the efficiency of the emitters will be drastically reduced.

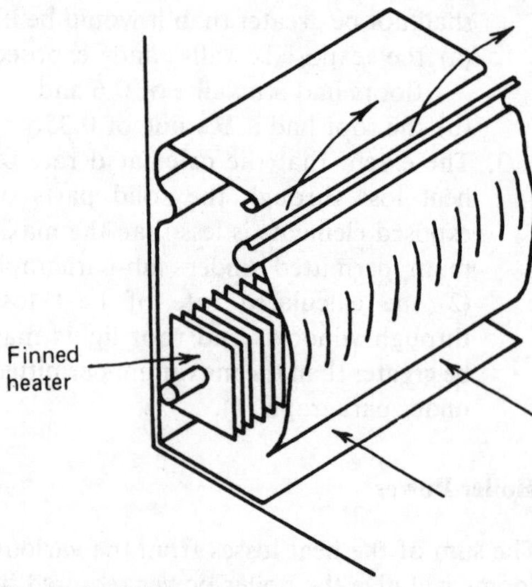

Fig. 12.8 Skirting convector.

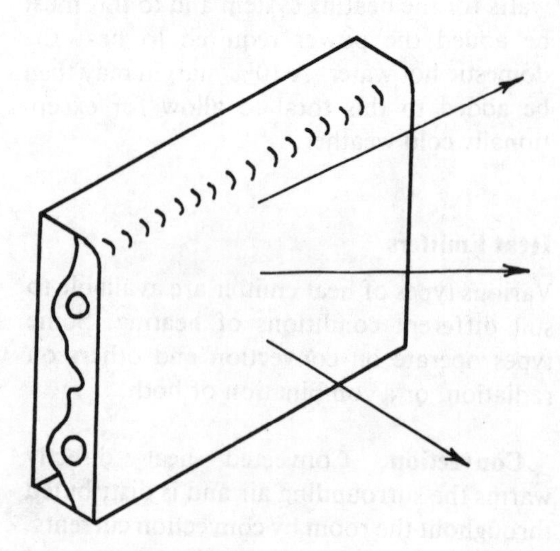

Fig. 12.9 Radiant skirting heater.

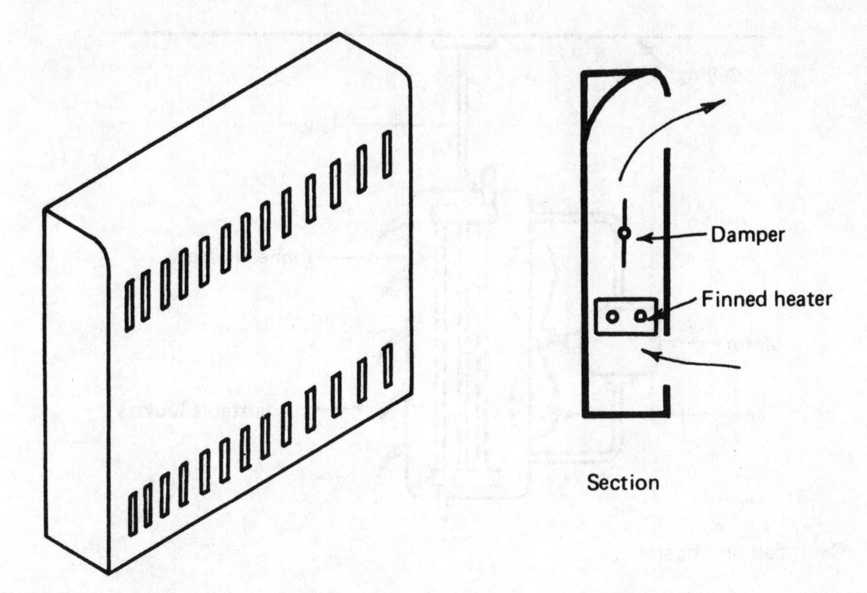

Fig. 12.10 Natural convector.

Radiant Skirting Heaters (Fig. 12.9). These help to reduce cold draughts at floor level whilst still providing radiant heat at higher level.

Natural Convectors (Fig. 12.10). These consist of a finned heater placed at low level inside a metal casing. If required, a damper may be incorporated inside the heater to control the flow of air through the emitter. A tall heater will transmit more convected heat than a short heater.

Fan Convectors (Fig. 12.11). These consist of a finned heater, usually placed in the top of a metal casing, with a fan or fans placed at the bottom, below the heater. The fan or fans may be two- or three-speed types and the

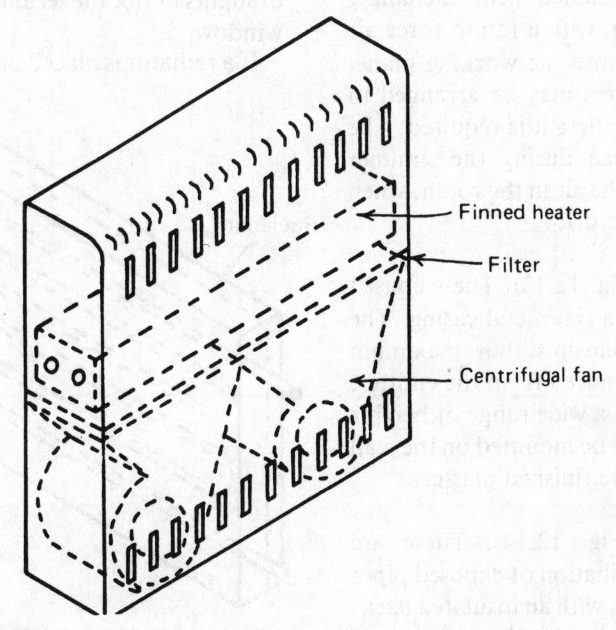

Fig. 12.11 Fan convector.

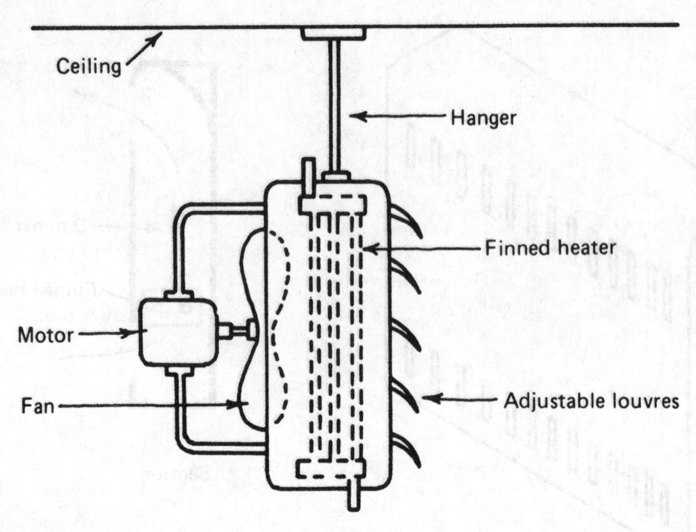

Fig. 12.12 Overhead unit heater.

control of the heat output from the emitter may be achieved by altering the speed of the fan or fans.

These emitters have the advantage over natural convectors of providing quicker heating of the air in the room and of providing better heat distribution. The additional cost of running the fan or fans must, however, be taken into account.

Overhead Unit Heaters (Fig. 12.12). These again consist of a finned heat exchanger inside a metal casing, with a fan to force air through the heater onto the working plane. The adjustable louvres may be arranged to direct the warm air where it is required. The emitters may be used during the summer months to circulate the air in the room, when the heater is switched off.

Radiant Panels (Fig. 12.13). These consist of a pipe coil inside a flat metal casing. The flat metal surface ensures that maximum radiant heat is transmitted from the emitter. They are available in a wide range of heights and lengths, and may be mounted on the wall or ceiling, flush to the finished plaster.

Radiant Strips (Fig. 12.14). These are formed from a combination of exposed pipes and sheet metal plates with an insulated back. They may be arranged in continuous runs of up to about 30 m and may be installed in workshops or warehouses.

Positions of Radiators

If a radiator is placed under a window, as shown in Fig. 12.15a, the rising stream of air mixes with the incoming cold air from the window. This provides a good distribution of warm air inside the room and reduces cold draughts at floor level and cold spots near the window.

If a radiator is placed on the wall opposite a

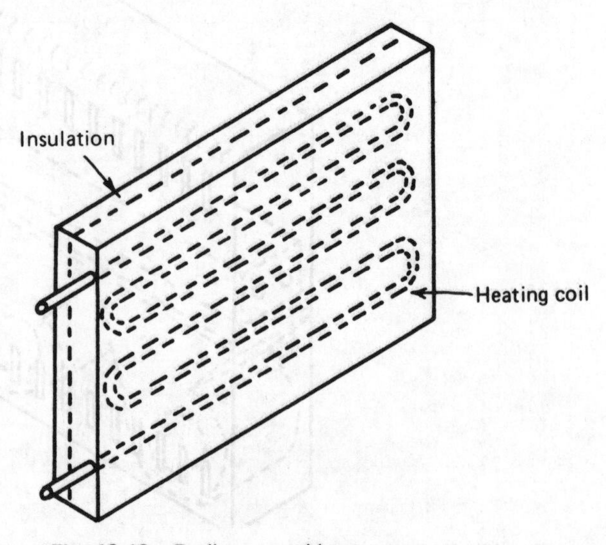

Fig. 12.13 Radiant panel heater.

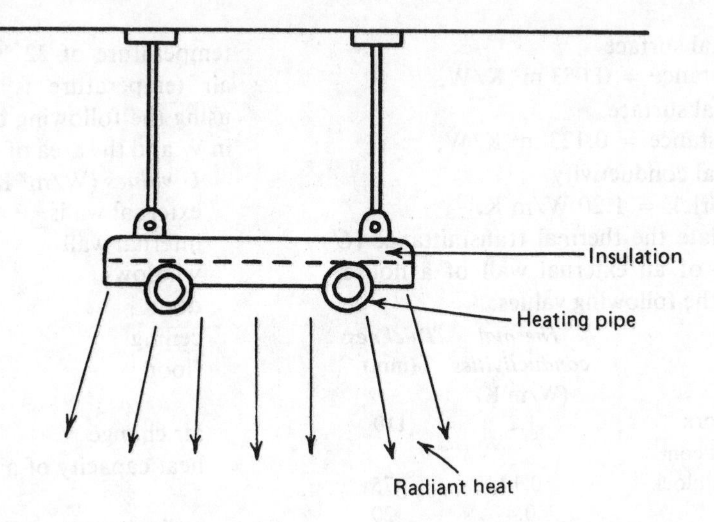

Fig. 12.14 Radiant strip.

window, as shown in Fig. 12.15b, the cold air from the window will flow downwards and across the floor. This will cause cold draughts at floor level. Greater thermal comfort will therefore result by placing radiators below the windows.

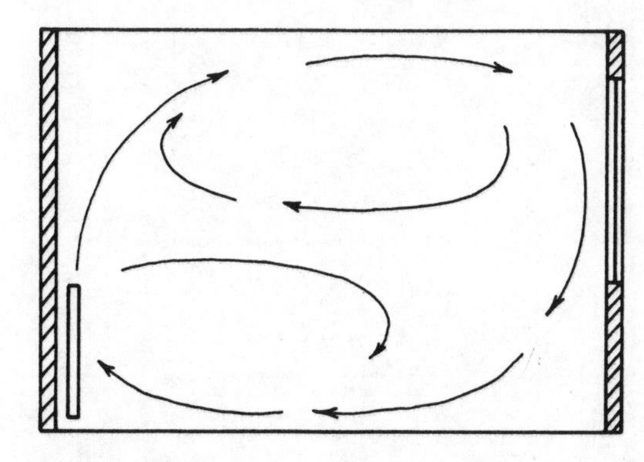

(a) Under the window

(b) On inside wall opposite the window

Fig. 12.15 Positions of radiators.

Painting of Radiators

The paint applied to the surfaces of radiators and panels does not effect their efficiency as convectors, but it can substantially reduce the amount of heat transmitted by radiation.

The best radiating surface at high temperature is matt black paint. Research by various authorities, however, has shown that the radiant heat emitted from various painted surfaces (gloss or matt), including green, red and white, is much the same below a temperature of 100 °C.

Metallic paints, such as aluminium, bronze or gold, may reduce radiation by as much as 90%, although convected heat remains unaffected.

Exercises

1. Define
   (a) thermal conductivity,
   (b) thermal resistivity,
   (c) thermal resistance,
   (d) thermal transmittance.
2. Calculate the thermal transmittance (U value) of a solid brick wall 105 mm thick, using the following values:

external surface
resistance = 0.053 m² K/W,
internal surface
resistance = 0.123 m² K/W,
thermal conductivity
of brick = 1.20 W/m K.

3. Calculate the thermal transmittance (*U* value) of an external wall of a house, using the following values:

	Thermal conductivities (W/m K)	*Thickness* (mm)
brickwork	1.2	110
aerated concrete block	0.14	75
plaster	0.4	20

Thermal resistances

inside surface	0.120 m² K/W
outside surface	0.053 m² K/W
cavity	0.180 m² K/W

4. Calculate the heat required to ventilate a room measuring 6 m long × 3 m wide × 2.5 m high, when the air changes are three per hour and the inside and outside air temperatures are 20 °C and −1 °C respectively.
5. A room measuring 5 m long × 4 m wide × 2.5 m high has two windows each measuring 1.8 m wide × 1.2 m high and a door measuring 1 m wide × 2 m high. A room adjoining one of the long sides will be heated to 18 °C and the other three sides are to be external walls. The windows are to be in the external walls and the door in the internal wall.

   The room is to be heated to a temperature of 22 °C when the outside air temperature is −1 °C. Calculate, using the following data, the heat losses in W and the area of the radiators in m².

U values (W/m² K)	
external walls	0.96
internal wall	2.25
windows	2.80
door	2.30
ceiling	0.90
floor	0.36
air change	$1\frac{1}{2}$ per hour
heat capacity of air	1340 J/m³K
radiator heat emission	540 W/m²

6. Explain how a room is heated (a) by convection and (b) by radiation, and state the advantages of these methods of heat transfer.
7. Sketch the following types of heat emitter and explain their operation:
   (a) skirting convector,
   (b) natural convector,
   (c) fan convector,
   (d) radiant panel,
   (e) overhead unit heater.
8. Explain why a heat emitter sited below a window is preferable to a heat emitter sited on an inside wall opposite a window.
9. What effect would the painting of radiators, by various types and colours of paint, have on the heat transmission from the radiators?

Chapter 13
Sealed Heating Systems

Principles

In any water heating system, provision must be made for the expansion of water. In the open system described in Chapter 11, an expansion and feed cistern is required. This cistern requires a ballvalve and overflow pipe and must be large enough to accommodate the expanded water plus a margin of $33\frac{1}{3}\%$ for extreme cold weather, when the water temperature will be higher and expansion greater.

The open system also requires an open vent and cold feed pipes, so care has to be taken in the location of the pump, or water may be forced out of the vent pipe or air drawn through it. If air is drawn into the system through the vent pipes, air locks may result and there is a greater risk of corrosion of the boiler and radiators.

The water in the cistern is also in contact with the atmospheric air, which results in its evaporation and entry of air into the system, which again may cause corrosion in the system.

Flexible-Membrane Expansion Vessel (Fig. 13.1)

In the sealed system of heating, the open expansion and feed cistern is replaced by a flexible-membrane expansion vessel. The vessel is fitted near the boiler, either on the flow or return pipe. The vessel may also be piped away from the boiler by the following pipe sizes:

up to 24 kW, 15 mm diameter
above 24 kW, 22 mm diameter

The vessel is divided into two compartments by means of a rubber diaphragm. One side of the diaphragm is filled with nitrogen gas and the other is connected to the heating system and thus filled with water. To support the static head produced by the height of the water in the system, the nitrogen gas is pressurised so as to produce a minimum water pressure at the highest point on the circuit of 10 kPa.

Size of Expansion Vessel

The expansion vessel must be sized so that its volume is at least equal to the volume of expansion of water in the system, when raised from 10 °C to 110 °C, irrespective of the normal working temperature. Table 13.1 may be used for calculating the approximate size of the vessel.

System Characteristics

The system should conform to the *Heating and Ventilating Contractors Association Guide to Good Practice; Smallbore and Microbore Domestic Central Heating, Part 2, Sealed Systems*. The expansion vessel should meet the requirements of BS 4814: Expansion Vessels in Sealed Hot Water Systems.

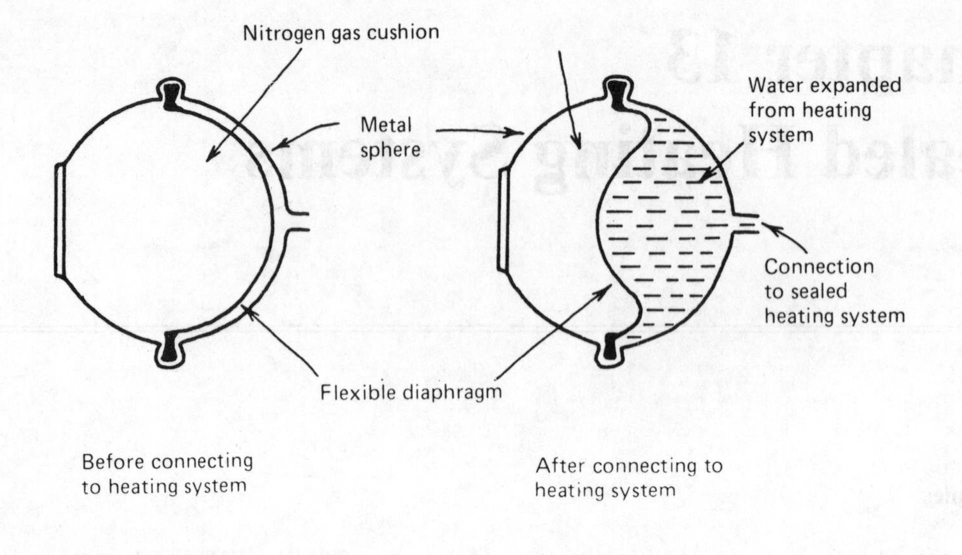

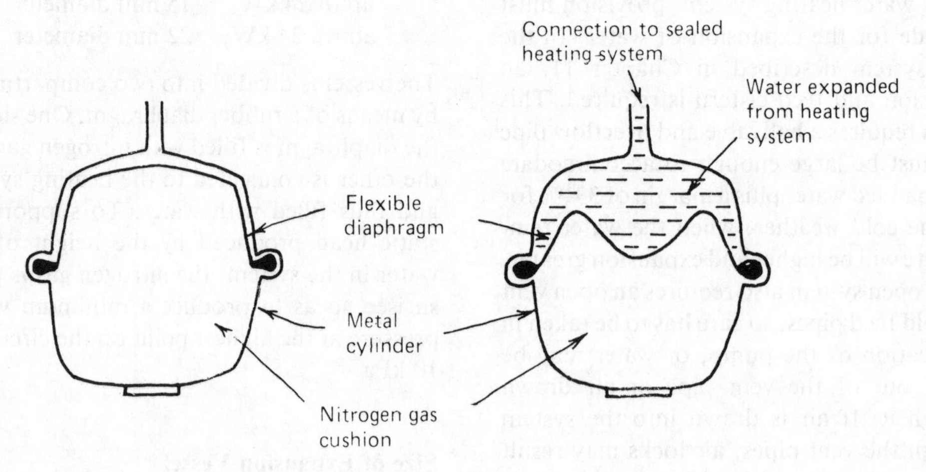

Fig. 13.1 Expansion vessels.

Table 13.1 Approximate Sizes of Expansion Vessels

System load (kW)	Nominal vessel size (l)	
	Traditional boiler and radiators	Low content boilers and fan convectors
3	4	4
6	4	4
12	8	4
18	12	8
24	18	8

The size of the system is limited to a heating load of 45 kW, but the general installation principles, given below, apply equally to much larger systems. The essentials of the HVCA Code and the British Standard are summarised as follows:

- Maximum flow temperature, 99 °C
- Temperature drop in the circuit, 8–17 °C
- Maximum boiler working pressure, 300 kPa
- Pressure relief valve blow off pressure 30 kPa above maximum working pressure
- Minimum pressure at highest point on the circuit, 10 kPa
- Maximum water velocity in the circuit, 1.5 m/s

Safety Devices

1. Pressure relief valve with non-metallic disc and metal seat with discharge pipe in a visible but safe place where there is no risk of contact with the hot water by persons using the building. The discharge should be via an air break to a tun dish.
2. Automatic controls and limit devices, on the heat or energy input, so designed as to ensure that the water temperature is controlled so as not to exceed 100 °C.
3. Thermal relief valve set to open should the water temperature exceed 100 °C. The valve should be connected to a discharge pipe as described in part 1.
4. Pressure gauge and thermometer fixed to the boiler or connecting pipe.

Hot-Water Cylinder

The primary heating element of the cylinder should be rated at 35 kPa higher than the relief valve setting. The heat recovery time for the water when heated from 10 °C to 66 °C should be 2 h. If the primary circuit to the cylinder operates on natural convection, the diameter of the pipes should not be less than 22 mm.

Advantages of the Sealed System

Apart from the cost saving brought about by not having to install an expansion and feed cistern in the roof and provide all the necessary pipe connections, the system has the following merits.

1. Being a sealed system, there is less risk of oxygen entering the system.
2. Pumping over or drawing air through the vent pipe is eliminated.
3. Noise due to bubble formation in the boiler or gas-fired heat exchanger is eliminated because of the increased water pressure.
4. The boiler may operate at a higher temperature, thus reducing the areas of the radiators and diameters of pipes.
5. The boiler may be sited on the roof, thus reducing the length of the flue and saving valuable space at the lower level.

Layout of the System

Fig. 13.2 shows a system incorporating a combined expansion vessel, pressure gauge, automatic air valve and pressure-relief unit. The unit will save a good deal of installation time.

Fig.13.3 shows a combined heating and hot water supply system supplied with cold water from a top-up bottle. A non-return valve in the cold feed pipe prevents hot water flowing into the bottle from the heating system.

Fig. 13.4 shows a sealed micro-bore system with a rooftop boiler room. Several radiators may be connected to the manifolds by means of 6, 8, 10 or 12 mm bore soft copper pipes.

Heat Emitters

Radiators or convector heaters may be used for the system, but the temperature of any exposed part should be at a maximum of 82 °C, or there will be a danger to the occupants.

Filling and Commissioning the System

The system may be filled by a top-up bottle or direct from the water main using a flexible hose, providing a non-return valve and stop valve is incorporated. The system may also be filled by means of a force pump drawing water from a cistern. Sealed systems may require topping up with water in the first few months of operation.

In commissioning the system, it is generally advisable to maintain a low pressure and high temperature for a period of a few hours, to assist the elimination of air. When the system cools down, the system should be adjusted to

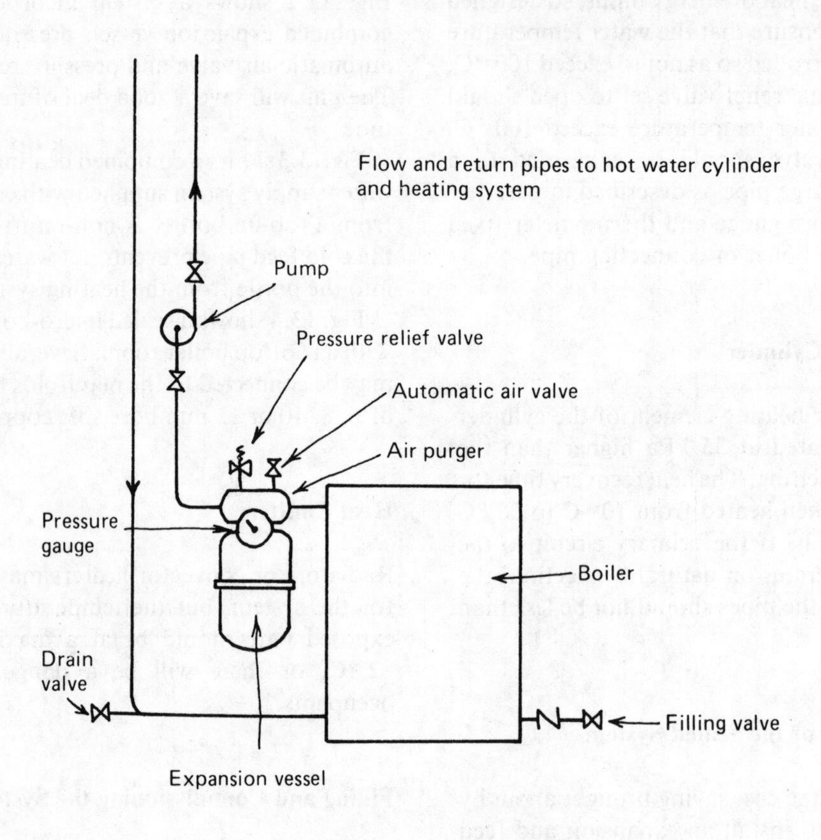

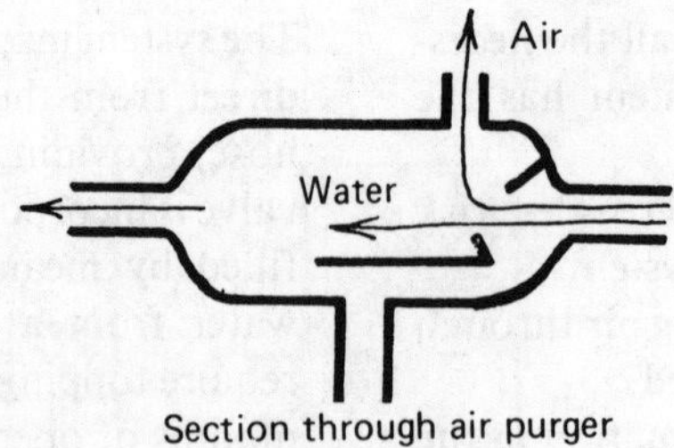

Fig. 13.2 Combined expansion vessel, pressure gauge, automatic air valve and pressure relief valve unit.

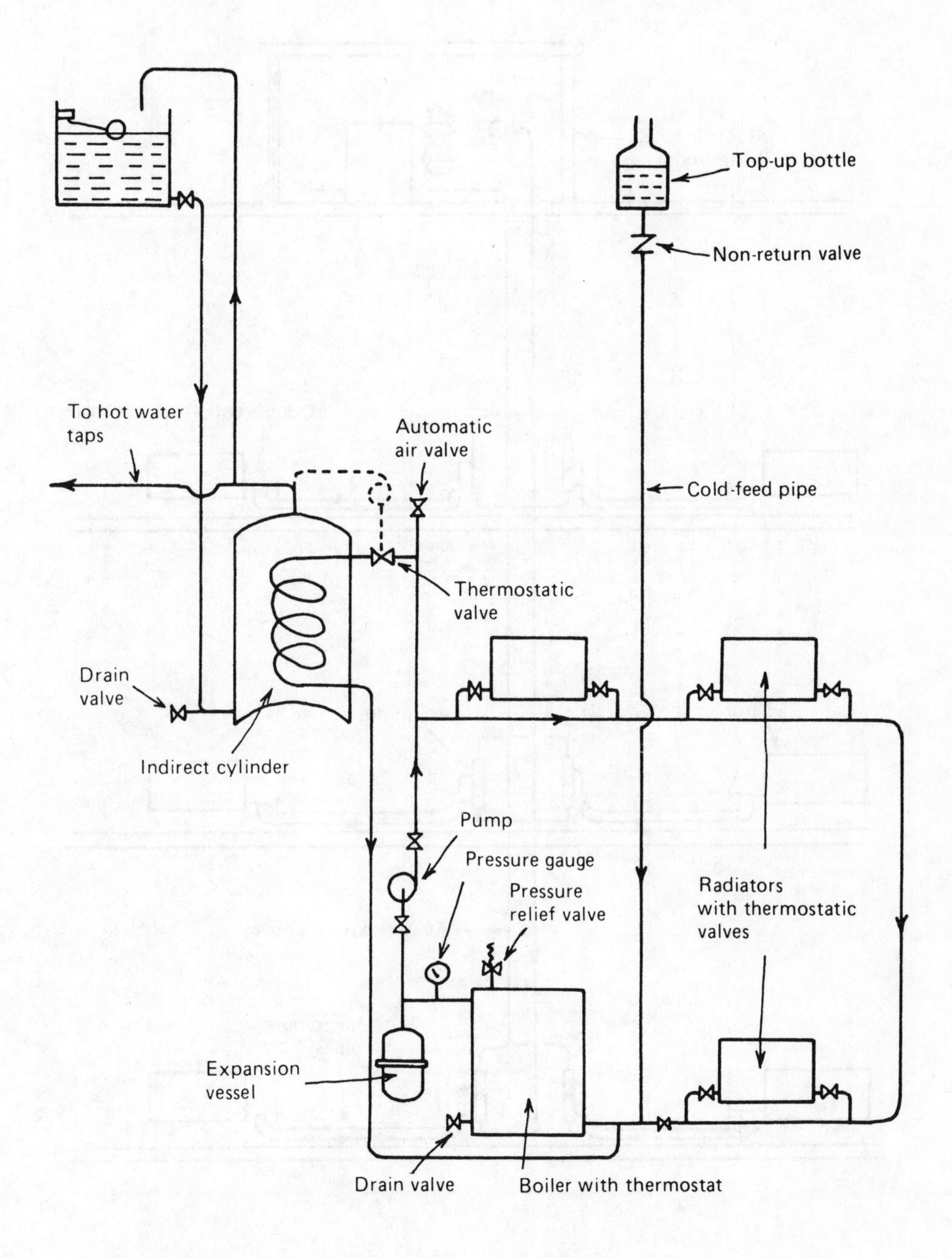

Fig. 13.3 Combined heating and hot-water supply sealed system.

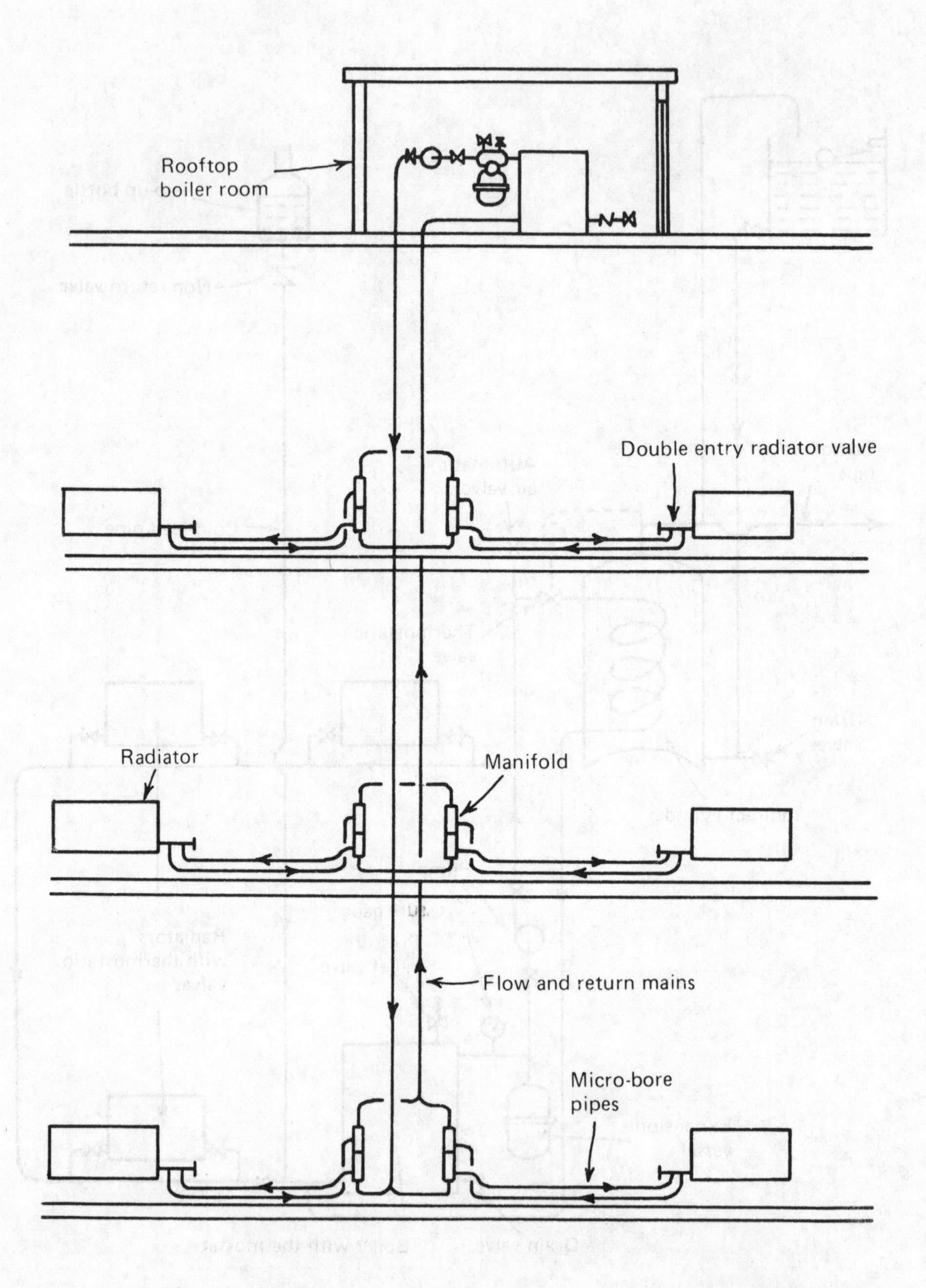

Fig. 13.4 Sealed micro-bore system with rooftop boiler room.

a value equal to the static head, plus a margin of 50 kPa. The final pressure at maximum temperature, and with all the heat emitters and circuits in operation, should not exceed 300 kPa.

Example 13.1. Calculate the pressure of gas in the expansion vessel when the static head above the vessel is 5 m and the water is cool.

Pressure required = (Head in metres × 9.81) + 50
= (5 × 9.81) + 50
= 99 kPa
(100 kPa approx.)

Location of the Expansion Vessel

In siting the expansion vessel, consideration should be given to its position in relation to the position of the pump. The vessel should never be connected to the positive side of the pump, or the pump pressure will act on the rubber diaphragm and be added to the system water pressure.

The vessel may be connected to either the flow or return pipe, but, because the vessel diaphragm is manufactured from rubber, connecting it to the return pipe would provide cooler water and thus increase the life of the rubber. Fig. 13.5 shows a small combined heating and hot water supply system with the expansion vessel connected to the return pipe. The three-way diverting valve will control the temperatures of both the hot water supply and the heating system.

For larger systems operating at higher temperatures, it is advisable to connect the expansion vessel to a separate pipe and site the vessel away from the boiler.

Fig. 13.6 shows a system with an expansion vessel fixed to the wall and connected to the boiler by a noncirculating pipe. To avoid thermal circulation in the pipe, a vertical anti-gravity loop should be included and any isolating valve fixed on the pipe should be locked to prevent accidently closing and disconnecting the boiler from the vessel.

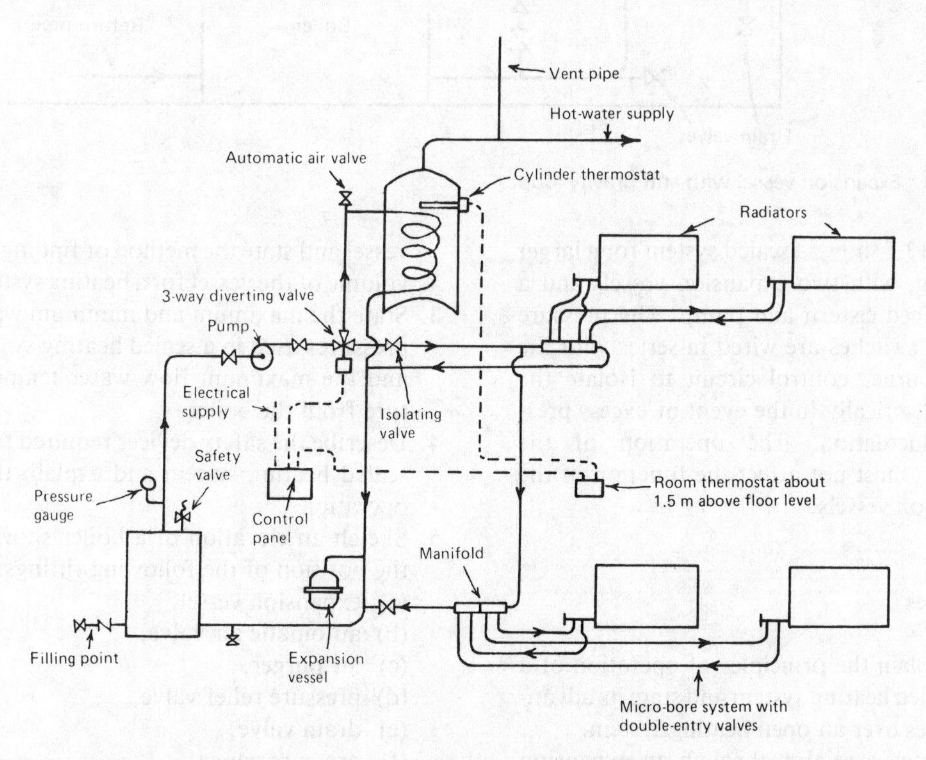

Fig. 13.5 Small sealed combined heating and hot-water supply system.

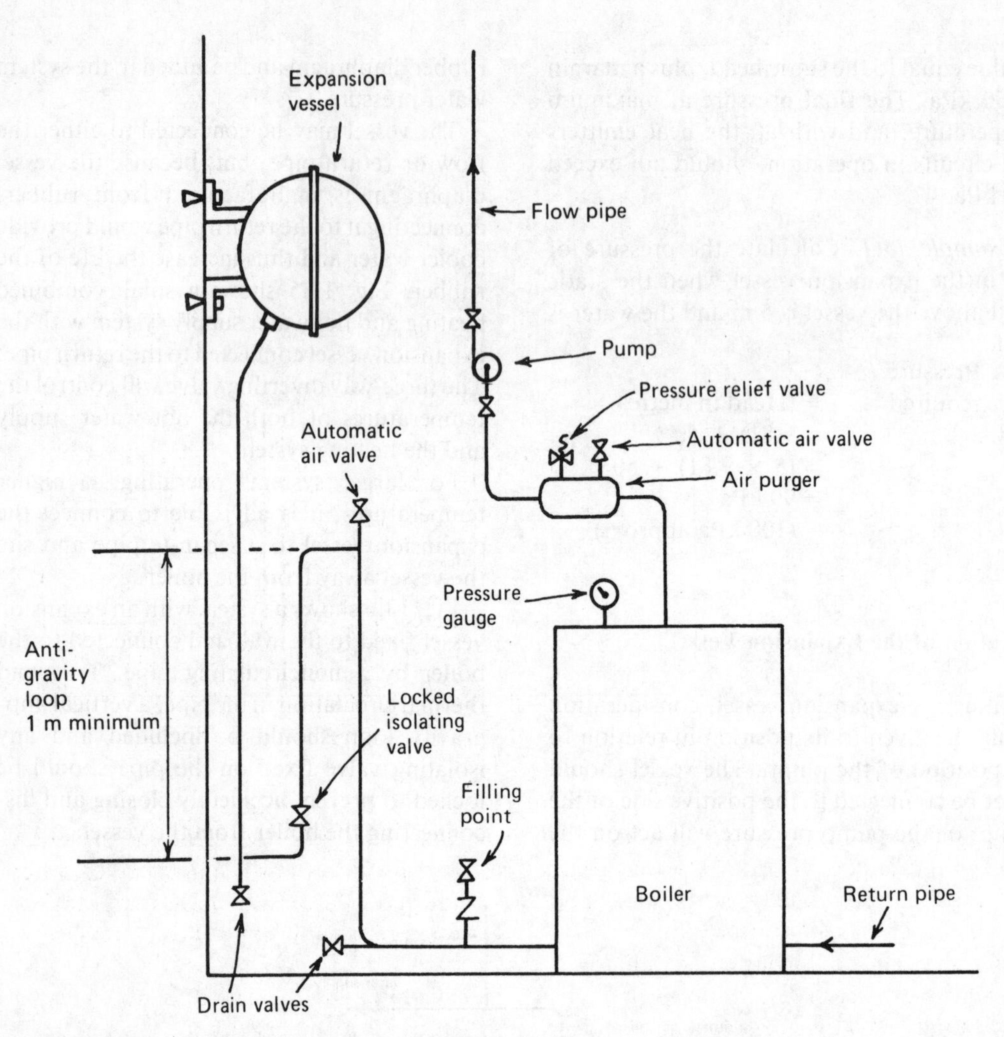

Fig. 13.6 Expansion vessel with anti-gravity loop.

Fig. 13.7 shows a sealed system for a larger building, with two expansion vessels and a boiler feed cistern and pump. The pressure cut-out switches are wired in series with the boiler-burner control circuit to isolate the plant electrically in the event of excess pressure fluctuation. The operation of the switches must not affect the function of the expansion vessels.

Exercises

1. Explain the principles of operation of a sealed heating system and state its advantages over an open heating system.
2. Sketch a section through an expansion vessel and state the method of finding the volume of the vessel for a heating system.
3. State the maximum and minimum water pressures used in a sealed heating system and the maximum flow water temperature from the boiler.
4. Describe the safety devices required for a sealed heating system and explain their operation.
5. Sketch an elevation of a boiler showing the position of the following fittings:
   (a) expansion vessel,
   (b) automatic air valve,
   (c) air purger,
   (d) pressure relief valve,
   (e) drain valve,
   (f) pressure gauge,

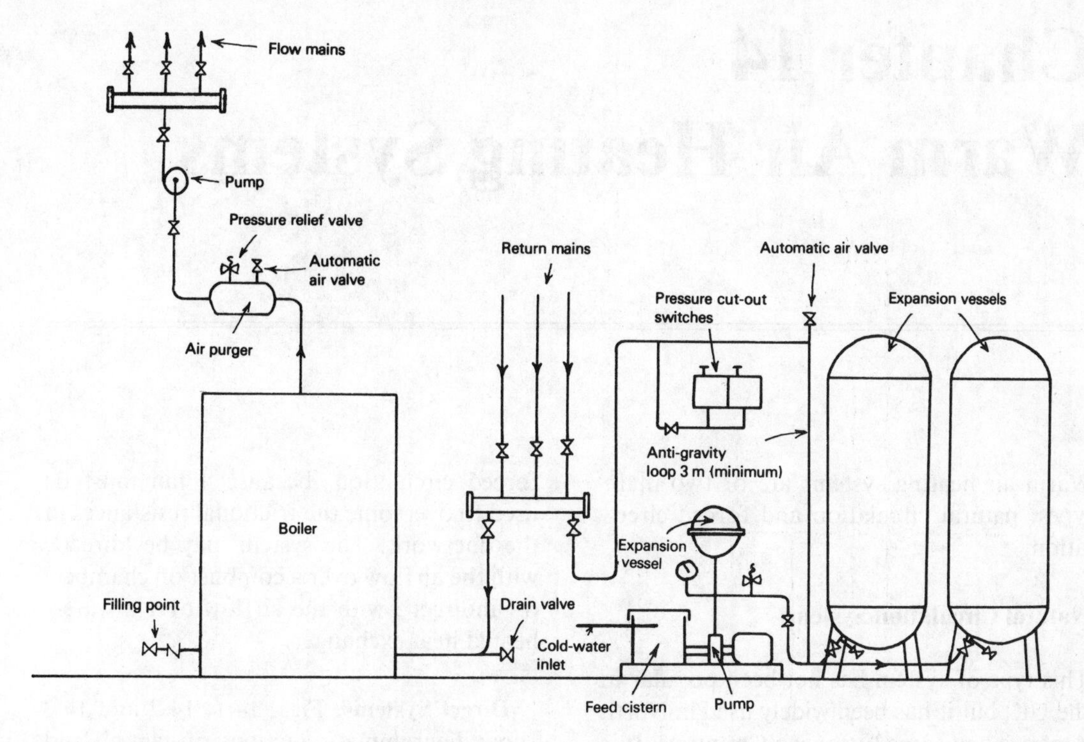

Fig. 13.7 Sealed system for larger building.

(g) filling valve.

6. Sketch an annotated isometric diagram of a sealed, combined heating and hot water supply system suitable for a three-bedroomed house having three radiators upstairs and downstairs respectively.
7. Sketch an elevation of a sealed heating system for a three-storey office block when the boiler is to be sited on the roof, showing all pipework and fittings.
8. Explain the operation of a top-up bottle in a sealed heating system.
9. Describe the method of filling and commissioning a sealed heating system.
10. Calculate the pressure of gas inside an expansion vessel when the head of water above the vessel is 8 m. A margin of 50 kPa above the static head pressure is required.
11. State the best location in the heating system for the expansion vessel and the reasons for using this position.
12. Sketch the installation of two large boilers used for a large sealed heating system showing expansion vessels, boiler feed cistern and pump. Explain the operation of the boiler plant.

Chapter 14
Warm Air Heating Systems

Warm air heating systems are of two main types: natural circulation and forced circulation.

Natural Circulation Systems

This type of system has not been popular in the UK, but it has been widely used in North America and some European Countries. It is generally suitable only for houses where there is a basement, so that the heated air can flow through the ductwork by natural convection created by the stack pressure. The air is heated by a furnace, which forms a direct heat exchanger, and this unit should be sited in a central position so that the duct lengths are reduced to a minimum.

Because the warm air flows through the ductwork by natural convection, the movement and pressure of the air as it enters the room is very small. The system, therefore, is easily affected by any positive pressure inside the rooms due to air infiltration from outside. One type of natural circulation employs a convector heater placed in a central position. The system is suitable for open-plan houses and bungalows and the heater may be fitted with a radiant panel to provide both radiant and convective heating. An advantage claimed for the natural circulation system is the absence of a fan, which uses electrical power and may give rise to noise problems.

Forced Circulation Systems

A more complex system of ductwork requires forced circulation, because a fan must be used to overcome the frictional resistances in the ductwork. The system may be 'direct', with the airflow over a combustion chamber, or 'indirect', with the airflow over a water-heated heat exchanger.

Direct Systems. Figs. 14.1, 14.2 and 14.3 show diagrammatic sections of gas, oil and solid-fuel direct air heaters. The gas and oil heaters required either a gas circulator or an electric immersion heater to heat the water in the cylinder. The direct system has advantages of:

1. very rapid heating up,
2. cost — it is usually cheaper,
3. lower resistance to airflow,
4. the outlet temperature may be higher,
5. no risk of freezing;

and disadvantages that

1. except for special types, they cannot provide hot water supply,
2. most units can be controlled only on the on/off or high/low principle,
3. because a flue is required, it is not always possible to site the unit in a central position.

Indirect Systems. Fig. 14.4 shows a diagrammatic section through an indirect air heater; the boiler also heats the hot water supply cylinder. An indirect system has the advantages that

1. it can be used with any type of boiler,
2. the boiler can also heat the water in the cylinder,

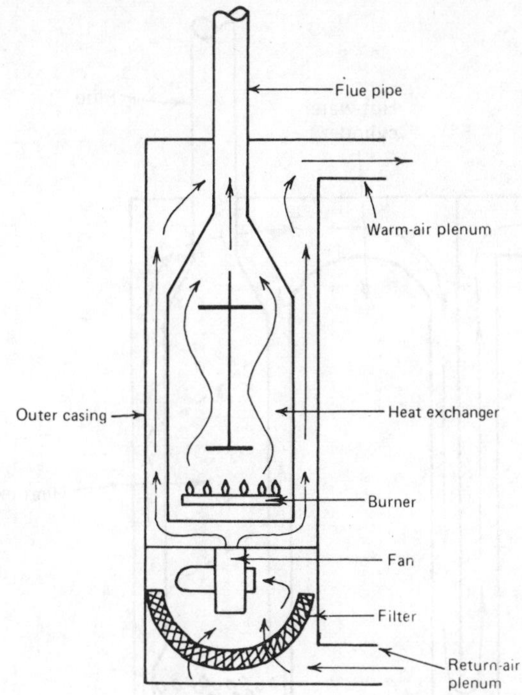

Fig. 14.1 Gas air heater.

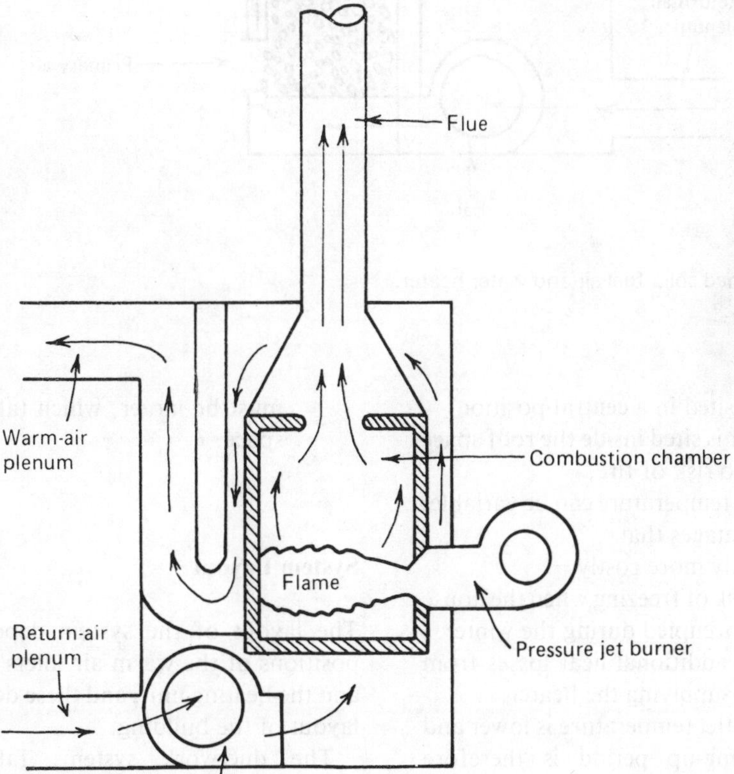

Flue
Warm-air plenum
Combustion chamber
Flame
Return-air plenum
Pressure jet burner
Fan

Fig. 14.2 Oil-fired air heater.

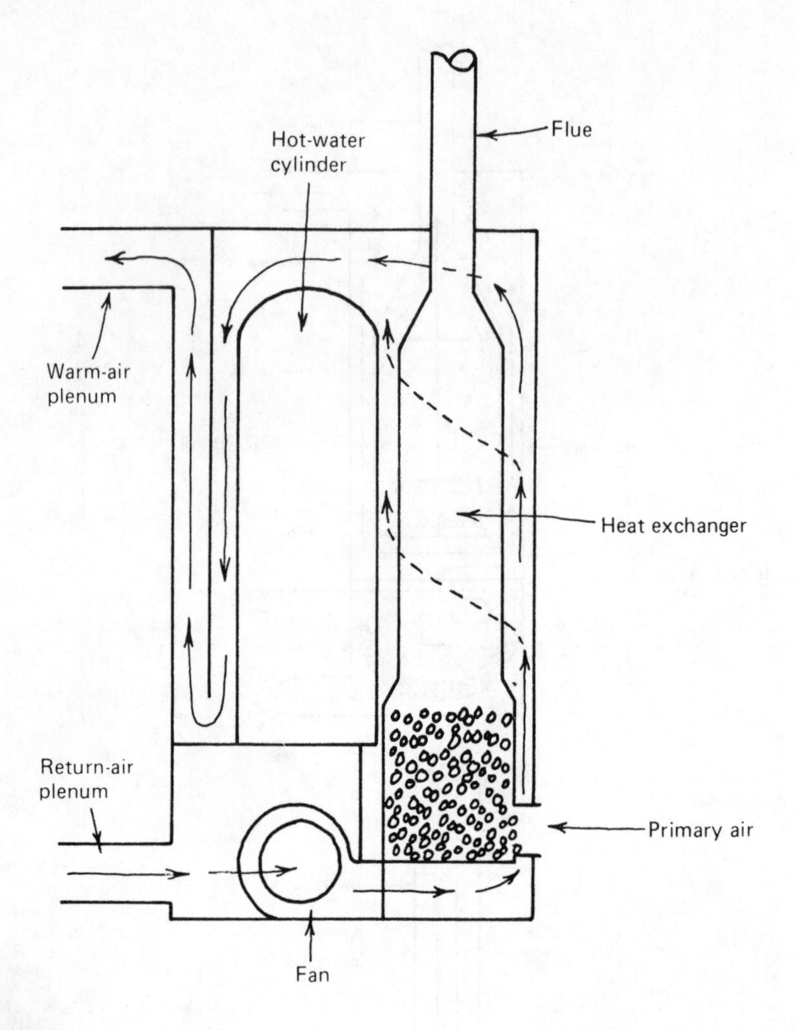

Fig. 14.3 Combined solid-fuel air and water heater.

3. it can be sited in a central position,
4. if the unit is sited inside the roof space there is no risk of fire,
5. the water temperature can be variable;

and the disadvantages that

1. it is usually more costly,
2. there is risk of freezing when the house is left unoccupied during the winter.
3. there are additional heat losses from the pipes supplying the heater,
4. the air outlet temperature is lower and the heating-up period is therefore longer,
5. because the temperature of the heat exchanger is lower, the heating unit must be larger, which takes up more space.

System Layout

The layout of the system depends on the positions of the warm air inlets and outlets, and the heating unit, and these depend on the layout of the building.

The ductwork systems fall into the following patterns. When the heater is sited in the centre of the building, an extended plenum or a radial system may be used (see

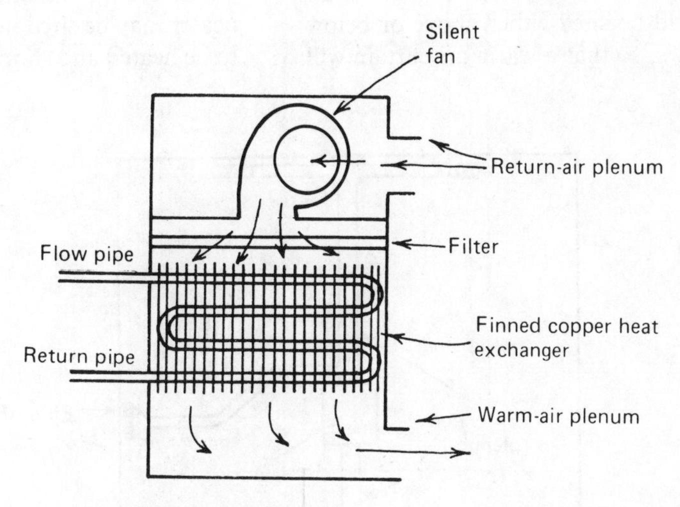

Section through heater

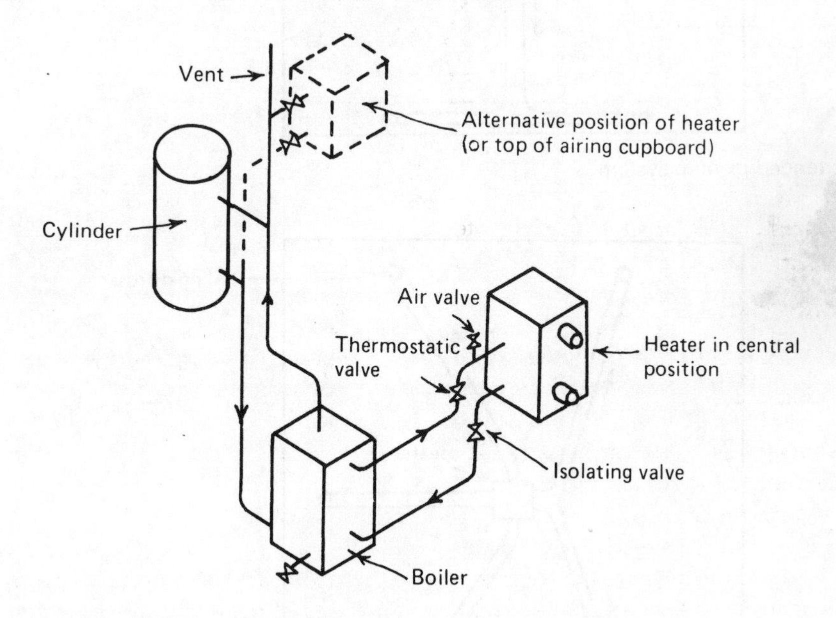

Fig. 14.4 Indirect air heater.

Figs. 14.5, 14.6 and 14.7). These systems ensure that the lengths of the individual ducts to the rooms are reduced to the minimum, thus minimising the frictional resistance. They also provide a better balance of air flow to each room. If the heater has to be sited at one end of the building, a stepped duct system is required as shown in Fig. 14.8. The air inlets should be sited either above or below the windows, so that a warm air curtain will flow past the windows and thus prevent the entering cold air from flowing along the floor.

Stub Duct System (Fig. 14.9)

To save on installation costs, the warm air heater may be sited in the centre of the room to be heated and short or stub ducts from the

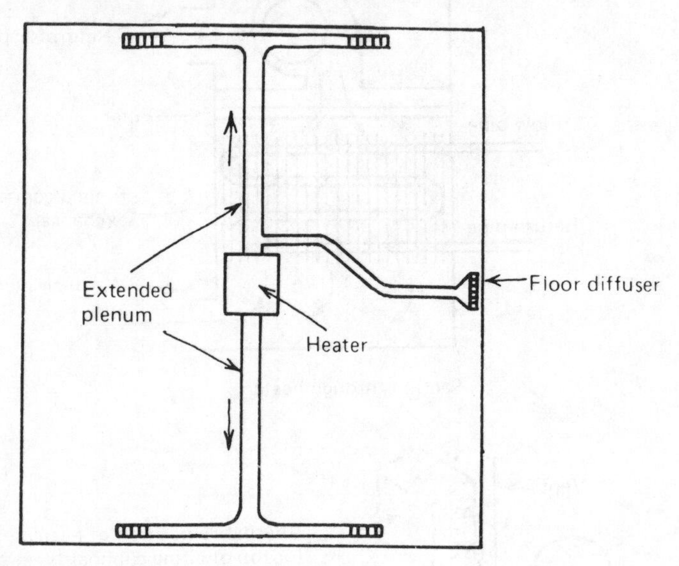

Fig. 14.5 Extended-plenum system.

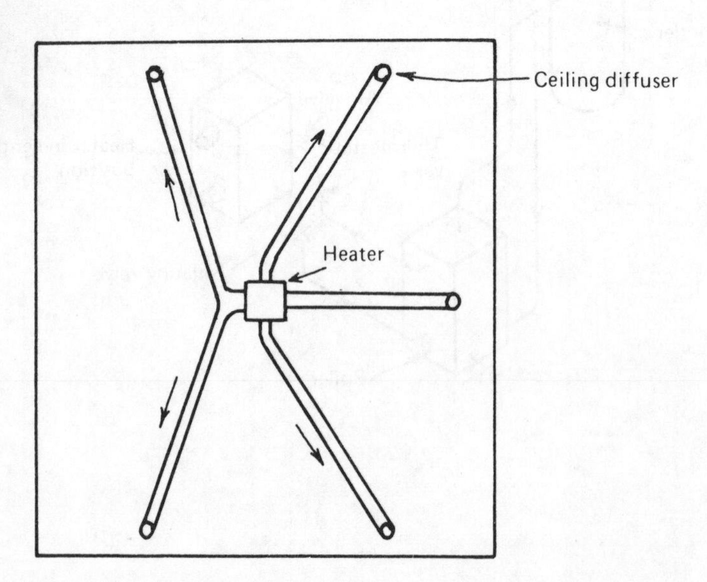

Fig. 14.6 Radial duct system with ceiling inlets.

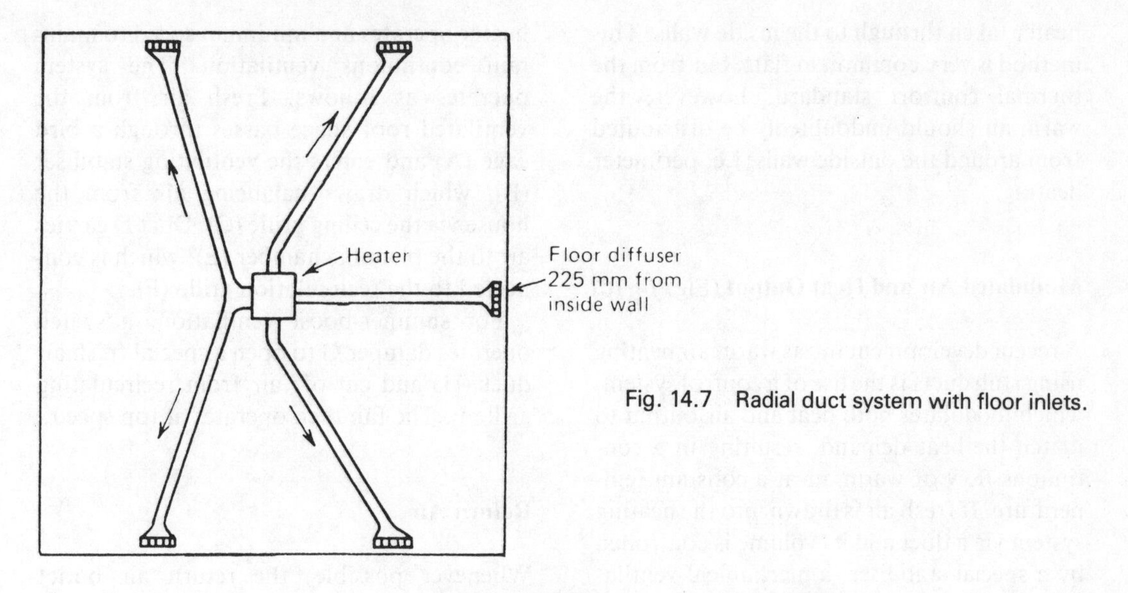

Fig. 14.7 Radial duct system with floor inlets.

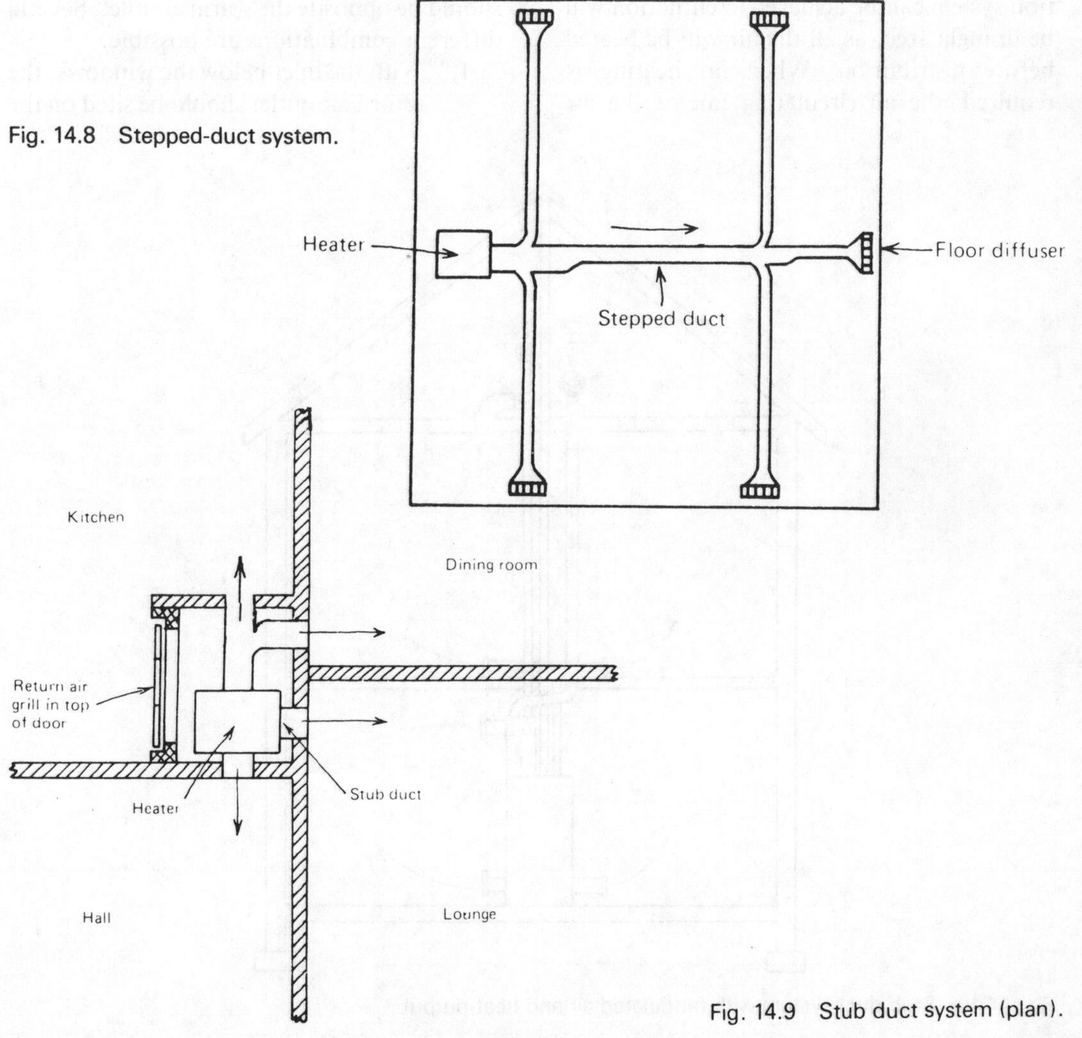

Fig. 14.8 Stepped-duct system.

Fig. 14.9 Stub duct system (plan).

heater taken through to the inside walls. This method is very common in flats, but from the thermal comfort standard, however, the warm air should undoubtedly be distributed from around the outside walls, i.e. perimeter heating.

Modulated Air and Heat Output (Fig. 14.10)

A recent development in gas warm air heating using stub ducts is the use of a control system, which modulates both heat and air output to match the heat demand, resulting in a continuous flow of warm air at a constant temperature. If fresh air is drawn into the heating system via a duct and its volume is controlled by a special stabiliser, a mechanical ventilation system can be achieved. Ventilation will be draught free, as all the air will be heated before distribution. When no heating is required, the air-circulating fan of the air heater operates at a minimum speed to maintain continuous ventilation. The system operates as follows. Fresh air from the ventilated roof space passes through a bird cage (A) and enters the ventilating stabiliser (B), which draws balancing air from the house via the ceiling grille (C). Duct D carries air to the plenum chamber (E), which is connected to the recirculation grille (F).

For summer-boost ventilation, a switch operates damper G to open a special fresh air duct (H) and cut off air from recirculating grille F. The fan then operates at top speed.

Return Air

Whenever possible, the return air outlet should be opposite the warm air inlet. Several different combinations are possible.

1. With the inlet below the windows, the return air outlet should be sited on the

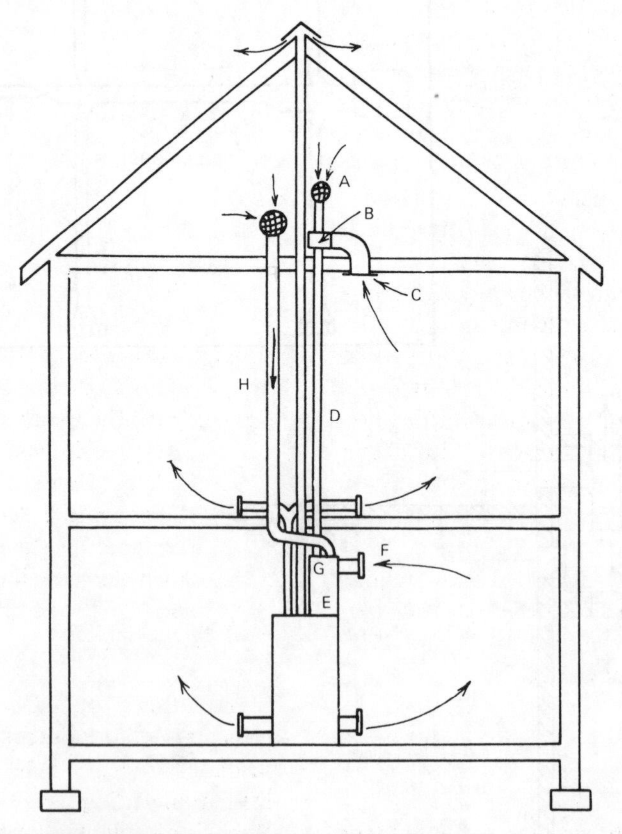

Fig. 14.10 Stub duct system with modulated air and heat output.

opposite wall just below the ceiling, or in the ceiling itself.

2. With the inlet above the windows, the return air outlet should be sited on the opposite wall just above the skirting level.

For domestic work, one main return air duct is usually sufficient and the duct can be carried through from the inside wall of the hall to the heater. The return air from the rooms is drawn back to the hall, where it passes through the duct to the heater. With this method, it is necessary to provide either sufficient space below the doors for the return air or a grille on the walls opposite the inlets.

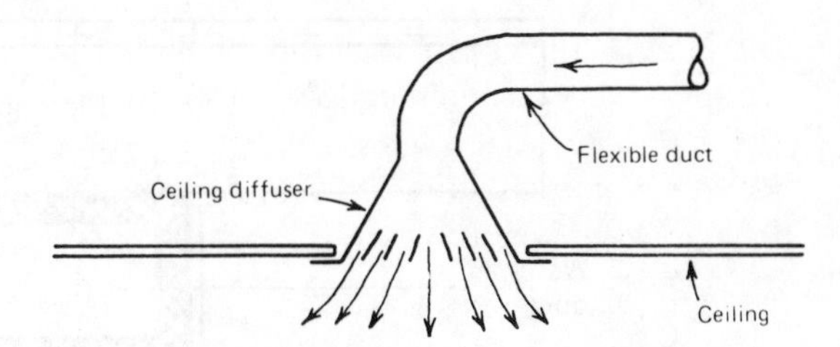

Fig. 14.11 Ceiling inlet.

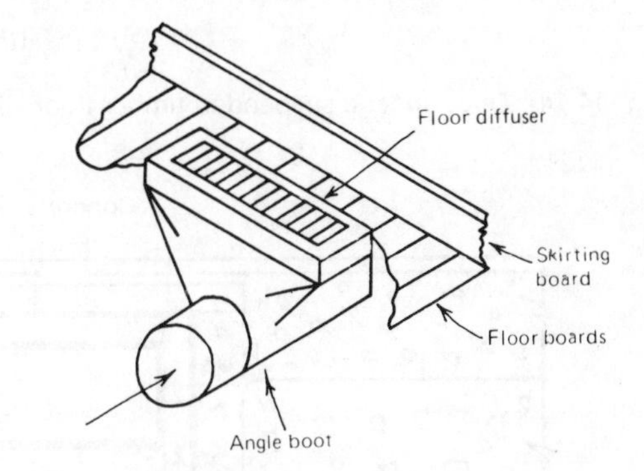

Fig. 14.12 Floor inlet.

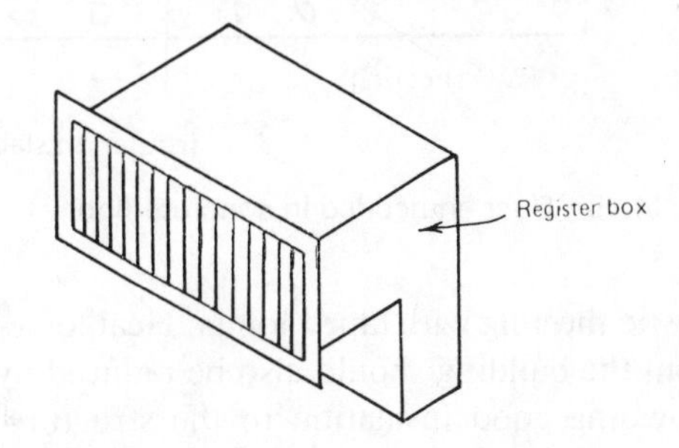

Fig. 14.13 Wall inlet.

Fresh Air

It is good practice to allow for some admixture of fresh air from the outside with the return air, and the fresh air duct should be large enough to provide about 25% of the total air requirement. Difficulty sometimes exist where, because of the wind, warm air is extracted instead of fresh air entering. This can be overcome by providing two fresh air inlets from opposite walls at right angles to each other, and bringing the two together in a 'balancing chamber' from which fresh air is taken into the return air duct. Another method of overcoming this problem is to fit a draught stabiliser on the fresh air duct.

In these days of ever-increasing fuel costs, fresh air admixture is sometimes omitted and fresh air brought in by natural infiltration through doors and windows. With either method, it is essential that there is always in adequate supply of air for combustion at the heater.

Figs. 14.11, 14.12 and 14.13 show ceiling, floor and wall inlets respectively. The same type of fitting show in Figs. 14.11 and 14.13 may also be used for outlets.

Ductwork

The ductwork may be circular, rectangular or square, and made from galvanised sheet steel or plastic. The following principles in designing the ductwork should be observed.

1. They should be as short as possible.
2. Fittings should be reduced to the minimum in order to reduce costs and frictional resistances.
3. All bends should have a large radius and T pieces should have swept branches.

Insulation of Ductwork. Wherever possible, all ductwork should be well insulated. Where there is a danger of the insulation getting wet, e.g. in a solid ground floor, it must be protected by wrapping with

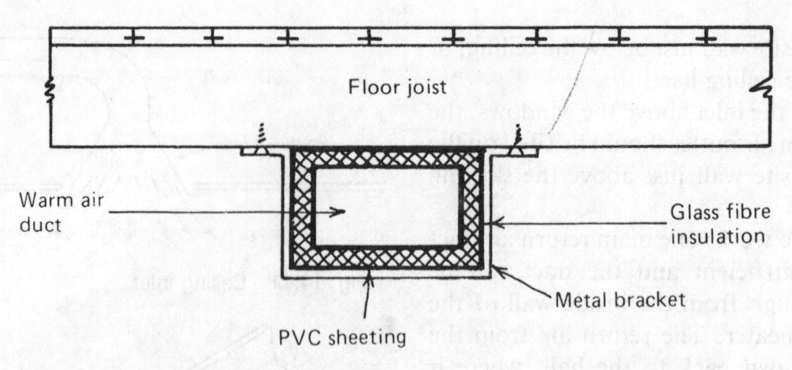

Fig. 14.14 Duct under a suspended timber floor.

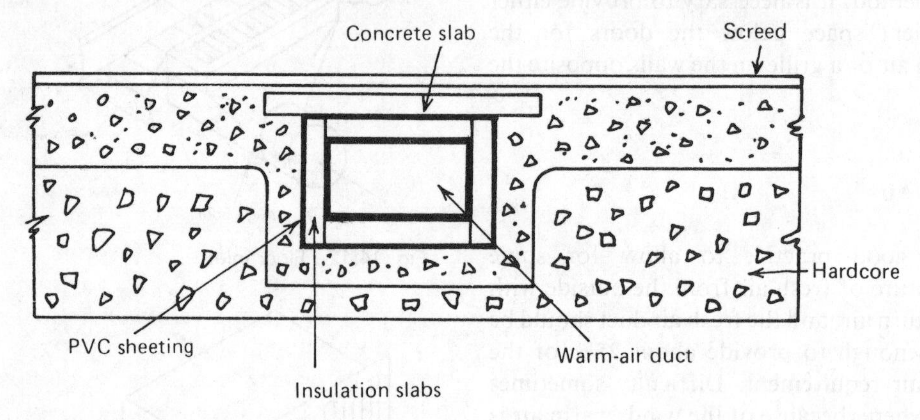

Fig. 14.15 Duct embedded in concrete floor.

plastic sheeting with taped joints. Heat losses from the building should also be reduced by providing good insulation to the structure. Figs. 14.14 and 14.15 show the methods of installing ductwork underneath a suspended timber ground floor and a solid ground floor respectively.

Sound Emission from Systems

Sound emission from warm air heating systems may be due to mechanical noise, vibration or air turbulence. Mechanical noise may be caused by hum of the electric motor, slack or dry bearings, or an unbalanced fan. Some manufacturers use resiliant mountings to insulate the fan and motor from the structural base. Where ductwork is connected directly to the heater or fan, a soundproof coupling in the form of an insulating rubber gasket or a short length of stout canvas should be provided. It is also necessary to limit the noise transmission in the vicinity of the heater by enclosing the heater with fireproof acoustic boarding. Noise from air turbulence is not usually a problem with smaller propeller-type fans and is therefore only likely to occur in systems having long lengths of ductwork which require centrifugal or axial flow fans. Noise of this type can be reduced by limiting the angular velocity of the fan and by using bends and fittings which change the direction of the air gradually.

Controls

Controls used in warm air heating systems fall into two groups:

1. controls such as dampers and adjustable grills which control the rate of air flow,

2. controls such as thermostats and time switches which control the air temperature and operating cycles. Thermostatically controlled motorised dampers may also be used. Controls are required to limit both high and low temperatures of air leaving the heater. The low temperature control is necessary to ensure that the fan does not start to circulate the air until it has been warmed up, or cold air will be forced into the rooms causing discomfort to the occupants.

Exercises

1. Describe natural and forced warm air heating systems and state the advantages of each system.
2. Describe direct and indirect warm air heating systems and state the advantages of each system.
3. Sketch an annotated diagram of an indirect warm air heating system for a bungalow.
4. Sketch vertical sections through gas, oil and solid-fuel air heaters and explain their operating principles.
5. Sketch the layout of the ductwork for the following warm air heating systems and state the advantages of each:
   (a) extended plenum,
   (b) radial,
   (c) stepped duct,
   (d) stub duct.
6. Show by means of sketches the method of installing warm air ducts
   (a) below a concrete ground floor,
   (b) below a timber floor.
7. Sketch and explain the operating principles of a modulated air and heat output warm air heating system and state its advantages.
8. Describe how the return air is brought back to the war air heater and also how some fresh air from outside is mixed with the return air.
9. State the materials that can be used for warm air ducts and the precautions which must be taken to reduce air flow resistance through the ductwork.
10. State the possible causes of noise from a warm air heating system and the methods that can be used for its prevention.

Answers to Numerical Exercises

Chapter 1

11. 2.943 mbar, 294.3 Pa

Chapter 5

12. 1500 l, 30 kW
13. 74 mm (use a 76 mm diameter pipe)
14. 9 m (approx.)
15. 16
16. 31.3 (use a 32 mm diameter pipe)

Chapter 8

2. 3.8 W/m^2 K
3. $U = 0.54$ W/m^2 K (approx.)
4. 935.6 W (approx.)
5. 2268.5 W and 4.2 m^2

Chapter 9

10. 128.5 kPa (approx.)

Index